Real Time Modeling, Simulatio
of Dynamical Systems

Asif Mahmood Mughal

Real Time Modeling, Simulation and Control of Dynamical Systems

Asif Mahmood Mughal
Biomedical Engineering Program
Higher Education Commission
Member, National Curriulum Review Committe
Islamabad, Pakistan

ISBN 978-3-319-81629-6 ISBN 978-3-319-33906-1 (eBook)
DOI 10.1007/978-3-319-33906-1

Softcover reprint of the hardcover 1st edition 2016

Printed on acid-free paper

This Springer imprint is published by Springer Nature
The registered company is Springer International Publishing AG Switzerland

Dedicated to
Qaiser Yasin Khan

Preface

Modeling and simulation of dynamical systems has been taught all around the globe with many methods, including standard modeling techniques of differential equations. These differential equation models are later transformed into frequency domain for analysis. In the latter part of the last century, time domain analysis began to be taught at undergraduate and graduate levels. Many books are currently serving the purpose of introducing modeling and simulation of linear time invariant systems. In the current scenario, new degrees are evolving such as systems engineering, mechatronics engineering, and biomedical engineering, where concepts of different conventional degrees (electrical, mechanical, etc.) merge together. Students are always curious in taking elective courses which widen their depth of knowledge. Accordingly, the courses and relevant text are being introduced continuously for meeting this demand. A modeling technique that covers the interdisciplinary modeling problems in a much better way than any other technique is the "bond graph method." This scheme has been taught since the mid-1970s, and the best available text is *System Dynamics: Modeling, Simulation, and Control of Mechatronic Systems* by Dean C. Karnopp, Donald L. Margolis, and Ronald C. Rosenberg, now in its fifth edition. Most of the literature or lectures are drawn from this book, as all respectable authors themselves were students of Prof. H. M. Paynter, who introduced bond graph techniques in the late 1960s. Recent developments, especially software in modeling and toolboxes in MATLAB as well as Maple, allow students to use different modeling paradigms interchangeably. In addition, a modeling software 20-Sim originally used for bond graph now also provides other modeling techniques as well. A simple course only on bond graph modeling may interest students from different backgrounds. The purpose of this book is to introduce modeling, simulation, and control perspectives to mixed-background students of different undergraduate degrees at final (senior) year or as a first/elementary course at graduate level. This book introduces state space modeling derived from two modeling techniques, namely, Lagrangian formulation and bond graph method, followed by analysis of system and usage of model in control system design. This is a one-semester text derived into eight chapters for lectures and other texts as case studies in different applications. I myself taught this course five times at the graduate level and improved the text accordingly in order to meet the requirements of students from various engineering disciplines and concentration areas. This book in no capacity is a replacement of *System Dynamics*, which is always the unique reference book for students and teachers in terms of examples, exercises, and descriptions of problems. I myself prefer that students should attempt the examples and problems of this book for better understanding and practice to get a good grip on the subject.

The book presented here, *Real Time Modeling, Simulation and Control of Dynamical Systems*, provides a better and comprehensive study text to cover over one semester. The focus is to introduce state space modeling and Lagrangian and bond graph techniques such that students can work on advanced problems easily after this course. The book also introduces MATLAB commands and most of the examples and problems are developed in 20-Sim software. Analysis and simulation chapters use MATLAB and Simulink in order to broaden students' knowledge of various software utilities. Real-time modeling means covering the system in differential equations or equivalently state space formulation. The analysis of the state space model is presented in time domain with concepts of eigenvalues and gains mainly. Frequency-based analysis is briefly discussed. The input-based responses are also simulated in time domain with state space and not with frequency domain. The present control techniques are well understood with pole placement rather than frequency domain-based analysis; hence the focus of time-based responses are given as compared to frequency-based responses, which are mostly popular in current textbooks. The chapters are organized systematically in the natural way so students keep on building from previous knowledge. There are no recommended prerequisites for engineering students as they already cover engineering mathematics in detail before taking these electives. However, universities that offer graduate-level engineering programs to non-engineering graduates should check for engineering mathematics knowledge before recommending this course or make an "advanced engineering mathematics" course as a prerequisite to it. The chapter descriptions are given as follows.

Chapter 1: The first chapter is very important in this text in order to explain details of any system, its dynamics, what a model is, and why to model. This chapter discusses in detail the model types and classifications of models in engineering practices. Furthermore, state space formulation is introduced from ordinary differential equations for nonlinear as well as linear time varying and invariant systems. Concepts of linearization and transfer function realizations are also discussed. The state space realization developed in this chapter will further be used in Lagrangian and bond graph techniques in the following chapters.

Chapter 2: Lagrangian modeling is discussed in this chapter from basic to advance level. A single-point rigid body model in Cartesian coordinates is initially discussed and converted into polar coordinates. Later the same model is expanded in vector notation for any generalized coordinates (Cartesian, spherical, cylindrical, etc.). The basic physics behind work-energy relationship is discussed with conservative and nonconservative forces. At the end of the chapter, state space formulation is obtained and elaborated with a few examples.

Chapter 3: This chapter introduces preliminary concepts of bond graph modeling techniques by drawing analogies of different energy systems. Concepts of power and energy variables, word bond graphs, causality, and flow of power are introduced with arguments building toward elements of bond graphs.

Chapter 4: Elements of bond graphs, which are 1-port, 2-port, and junctions, are discussed with their causality assignments and the governing or constitutive laws and examples of different energy systems. Examples within different energy systems with corresponding analogy are described. Junctions are discussed with signs of power flow as well as effort/flow equations. Modulated transformer and gyrator (2-port elements) are discussed in text, and modulated sources (1-port) are introduced in the problem section. It is assumed that students are well versed in basic knowledge of physics for electrical, mechanical, translational, and rotational systems.

Chapter 5: Writing analytical equations in state space formulation using constitutive laws is explained with different examples of electrical, mechanical, and mechatronics systems. As per existing literature, a generalized rule for all kinds of systems is not available. A simplified procedure for all kinds of systems is to be evolved into a basic methodology of bond graph as independent energy systems (electrical, mechanical, hydraulics, etc.) must not be violated. This procedure is emphasized and discussed to develop bond graph of multidisciplinary energy systems. This chapter develops capability in students to fully model different systems in bond graph and write its equations.

Chapter 6: Different concepts of advance bond graph modeling are summarized and combined in this chapter in a logical sequence so the reader can understand why specific advancement or extra elements are needed. It covers algebraic loops, derivative causality, and fields, as well as some advanced mixed system with partial bond graph portions. A further combination of elements in different requirements is discussed, and, at the end, the topic of vector bond graph is touched. This chapter helps to develop a significant grasp on the bond graphs method by exploring different key concepts so students can utilize them in their projects. This is the last of four chapters covering specific bond graph techniques.

Chapter 7: Analysis and simulation of state space systems are discussed with details in time domain and some analysis in frequency domain. Most engineering students are familiar with responses in transfer function approach, and in this chapter, these responses are discussed with state space method. Generalized solutions of state space equations followed by free and forced responses are discussed as homogenous and non-homogenous solutions. Concepts of internal and external stabilities are explained with free and forced responses, thus discussing damped and amplified systems. Forced responses are studied with impulse, step, sinusoidal, and decaying sinusoidal signals in order to elaborate the bounded input and bounded output of control system designs with stability planes and resonance.

Chapter 8: A complete paradigm of control system design based upon state space model is discussed in this chapter. It introduces a detailed simulation diagram for control system design followed by its simplification based upon design requirements. Open- and closed-loop observers are also discussed with complete compensator design. Performance parameters and tuning of PID and control gains are also discussed with the help of MATLAB tuning and constraint optimization blocks, which help to understand the requirement of control design. The basic idea is to familiarize the readers with different types of control systems, keeping the design scope for advance courses.

Chapter 9: This chapter consists of different projects and research work for multidisciplinary energy systems with special focus on bioengineering systems. Recent advances in bond graph allows the research in the area of biomedical and bio-mechatronics systems. Bond graph models of biomechanics systems are discussed briefly in order to introduce their applications in this field. The selected projects of this part can also be taught in the classroom for description, analysis, or assessment purposes depending upon the time of the semester.

The book is written for a semester workload in real-time modeling, simulation, and control system design. This can serve as a beginning course for control system specialization and may be taken with linear system theory during the same semester. I prefer and teach this course with a semester project requiring students to develop and simulate a model of any electromechanical or mechatronics system by utilizing bond graph technique. It is highly recommended that the book *System Dynamics* should be employed as a reference book and guideline for the semester project from students. The lecture-wise breakdown is given in the table below for a 16-week semester for a 3-credit-hour course.

Content	Duration (weeks)
Chapter 1: Dynamical Systems and Modeling	1
Chapter 2: Lagrangian Modeling	2
Chapter 3: Introduction to Bond Graph Modeling	1
Chapter 4: Elements of Bond Graph	1
Chapter 5: Analytical Formulation by Bond Graph Modeling	2
Chapter 6: Advance Bond Graph Modeling	2
Chapter 7: Simulation and Analysis of State Space Systems	2
Chapter 8: Introduction to Control Systems	2
Chapter 9: Recent Applications of Bond Graph Modeling	1.5
Assessments, Review, and Project Presentations	*1.5*

Islamabad, Pakistan Asif Mahmood Mughal

Acknowledgments

I am grateful to my Allah (أﷲ) (the Lord of the heavens and earth) for His countless blessings in my life. I begin with Salat-o-Salam (peace and blessing) upon His final messenger and prophet Muhammad (حضرت محمد صلى الله عليه و على اله و صحبه و امته وبارك وسلم تسليماً كثيرا) accepting that I am not capable of achieving any objective but with the blessings of Allah (أﷲ), the Great and the Majestic.

I dedicated this book to Mr. Qaiser Yasin Khan, who was my supervisor at work in my early days of professional life. He became a true source of inspiration for me during these years due to his excellent mental acumen, professional integrity, and visionary guidance. This dedication is a small appreciation for a very few qualities I tried to imitate from his legendary professionalism.

I am very thankful to Prof. Dr. Afzaal Malik, who introduced me to bond graph modeling during a graduate course which developed my interest in this area and lead to several publications. I am also thankful to D. C. Karnopp, D. L. Margolis, and R. C. Rosenberg, who responded to my queries and even sent me scanned explanations of handwritten and drawn solutions. I am highly indebted to my Ph.D. supervisor Dr. Kamran Iqbal from the Dept. of Systems Engineering, University of Arkansas at Little Rock (UALR). He taught me several subjects of controls and let me introduce bond graph modeling to his undergraduate course during my Ph.D. work. Any words of appreciation will be less for his contributions to enhance my skills, and he is always motivating me for new endeavors. I also like to especially acknowledge Tallal Saeed, Madiha Zoheb, Tayyaba Qaiser, Sana Javed, Maryam Iqbal, Hassan Javed, Muhammad Furqan, Mona Jaffar, and Mariam Javed, whose work became a significant contribution in this text. I am greatly thankful to Marta Moldvai for her precious suggestions and improvement of this manuscript and for providing me great encouragement and support during the review and publication process. I am beholden to my family that they all contributed in their capacity toward my academic and research accomplishments. I am unable to mention names of each and every person who helped or to acknowledge all supports to achieve this goal. I can only pray to Allah that He may bless all of them (*who helped me in any regard*) with what they need most in their life.

Asif Mahmood Mughal

Contents

1 Dynamical Systems and Modeling 1
1.1 State Variables and State Vector 2
1.2 Input and Output 3
1.3 Modeling 3
1.4 Analytical Modeling 4
1.5 Physical and Computational Systems 7
1.6 Mathematical Model 7
1.7 State Space Method 8
1.8 Linearization 10
1.9 Transfer Function 12
Problems 17
References 17

2 Lagrangian Modeling 19
2.1 Modeling in Three Axes 21
2.2 Modeling in Cartesian and Spherical/Polar Coordinates 22
2.3 Work and Energy Formulation 25
2.4 State Space Representation 28
Problems 32
References 33

3 Introduction to Bond Graph Modeling 35
3.1 Power Variables 35
3.2 Pseudo Bond Graph 37
3.3 Energy Variables 37
3.4 Excitation and Response 39
3.5 Word Bond Graph 39
3.6 Causality Assignment 40
3.7 Active Bond 40

3.8 Connection of Bonds . . . 43
Problems . . . 45
References . . . 45

4 **Elements of Bond Graph** . . . 47
4.1 1-Port Elements . . . 47
4.1.1 Sources . . . 47
4.1.2 1-Port Resistor . . . 48
4.1.3 1-Port Capacitor . . . 49
4.1.4 1-Port Inertia . . . 50
4.2 Derivative (Differential) Causality and Integral Causality . . . 51
4.3 2-Port Elements . . . 52
4.3.1 Transformer . . . 52
4.3.2 Gyrator . . . 54
4.4 Junctions . . . 54
4.4.1 0-Junction . . . 56
4.4.2 1-Junction . . . 56
4.5 Modulated Transformer and Gyrator . . . 58
4.6 Modulated Sources . . . 59
Problems . . . 59
References . . . 62

5 **Analytical Formulation by Bond Graph Modeling** . . . 63
5.1 Modeling Procedure . . . 63
5.2 Sensors . . . 70
5.3 Mechanical Systems . . . 70
5.4 State Space Equations . . . 74
5.5 20-Sim Software Tips . . . 82
Problems . . . 84
References . . . 84

6 **Advance Bond Graph Modeling** . . . 85
6.1 Algebraic Loops . . . 85
6.2 Derivative Causality . . . 88
6.3 Fields . . . 90
6.4 Series Motor . . . 93
6.5 Resistive Fields . . . 95
6.6 Bridge Circuit . . . 96
6.7 Combination of Elements . . . 96
6.8 Vector Bond Graphs . . . 97
Problems . . . 100
References . . . 102

7 **Simulation and Analysis of State Space Systems** . . . 103
7.1 Free Response of First-Order System . . . 105
7.2 Eigenvalues of Higher Order System . . . 107

7.3 Free Response of a Second-Order System 108
7.3.1 Undamped Systems 109
7.3.2 Damped Systems 111
7.4 Relationship Between Eigenvalue and Time 116
7.5 Damped Natural Frequency 116
7.6 Amplification Not Damping 118
7.7 Internal Stability of the System 121
7.8 Forced Response 123
7.8.1 Impulse Response 124
7.8.2 Step Response 127
7.8.3 Sinusoidal Response 128
7.9 Resonance 129
7.10 Decaying Sinusoidal Response 131
7.11 External Stability 134
7.12 Total Response of the System 137
Problems 139
References 141

8 Introduction to Control Systems 143
8.1 Representation of Controller 145
8.2 Error Model 146
8.3 Estimator or Observer 147
8.4 Control Design 150
8.4.1 Performance Specifications 150
8.5 Controlling the Response 152
8.5.1 PID Control 153
8.5.2 Compensation 154
8.5.3 Pole Placement Control Design 155
8.6 Shaping the Dynamic response 155
8.7 Gain Scheduling in Multivariable Control 157
Problems 158
References 159

9 Recent Applications of Bond Graph Modeling 161
9.1 Bond Graph Models of the Physiological Elements 161
9.1.1 Muscular Structures 162
9.1.2 Muscle Spindle Model 164
9.1.3 Golgi Tendon Organ Model 167
9.1.4 The Inertial Subsystem 167
9.2 The Complete Musculoskeletal System Model 168
9.2.1 The Extended Musculoskeletal Structure 171
9.3 Anthropomorphic Hand through Bond Graph 172
9.4 Movement Coordination Problem for Human Fingers 173
9.5 Electrical-Impedance Plethysmography 176
9.6 Evaporation in Infant Incubator 177

9.7 Customized Robotic Arm 178
9.8 Bilateral Master–Slave Telemanipulation 179
Problems 183
References 184

Index 185

Chapter 1
Dynamical Systems and Modeling

A system is an isolated part of the Universe that is of interest due to a specific reason for an appropriate time. This isolated reality can be used for discussion, study, analysis, improvement, betterment, protection, or any other objective. The isolated part in which we take interest actually becomes the ***system*** and everything else is surroundings, environment, or the rest of the Universe. For example, an astrophysicist wants to study the gravitational forces of the sun and its planets, so this isolated interest becomes a solar system for him. The rest of the galaxies, and even relatively small events on the earth and planets, become surroundings for him. An oceanographer analyzing the waves and currents in the Pacific is only interested in his isolated reality and the rest of the happenings in the world are collectively an environment. Similarly, an engineer working on a comparatively small problem whether designing, studying, or discussing is a system for him and everything else in the world becomes an environment or surroundings to him till he stops working. A car is a system, an isolated part of reality for a buyer. But this car itself is a combination of many other systems such as a combustion system, braking system, fuel injection system, electrical system, steering systems, etc. A buyer at a time is interested in all including the color of car, financial value in the market, fuel consumption, etc. He is overall looking for an appropriate driving system that suits his needs. But several different types of engineers are interested in several subsystems of the same car. This car in turn is a collection of subsystems. A subsystem is a complete system when one develops an interest in it and other subsystems of the same car become an environment or surroundings for it. This example is similar to the human body system with several subsystems like the respiratory system, digestive system, reproductive system, central nervous system, etc. A body needs all subsystems at a time but an orthopedic surgeon usually only focuses on the skeletal system, and a gastroenterologist is only interested in the digestive system.

Systems engineering approaches clearly distinguish between what the matter of interest is and what is not important in the current time. This leads to focus on achieving required objectives clearly and thoroughly. When boundaries are drawn,

A.M. Mughal, *Real Time Modeling, Simulation and Control of Dynamical Systems*,
DOI 10.1007/978-3-319-33906-1_1

then improvement of things inside boundaries can easily be targeted. The **Systems Engineering** approach leads to quantify the objectives and their solutions with well-established analytical methodologies. Modern day bioengineering is an example as it is solving a gap between myriads of experimental data and well-developed analytical theories. Generally the systems engineering approach quantifies the system's dynamics through mathematical representation in an appropriate time frame. A dynamical system is one that changes its state with time. In reality there is not a single system that exists in the entire universe that does not change its state. The system that appears static is either not considered within an appropriate time frame, or the change is out of the scope of interest. The North Star has been there for centuries and is responsible for helping with navigation. But the rotational axis of the earth changes during its movement in an elliptical path and about every 5000 years, the North Star passes his duties to another star for the same purpose. The appropriate time frame has to be defined properly when system dynamics are studied. Now let's consider an example of a book in a local library, which has never been issued and never been read. So in a library system the status of this book may be constant as it does not come into a scope of interest. But changes are there as pages are turning pale and binding is getting loose. For defining any **dynamical system,** two things are very critical: one is an appropriate time frame to notice its change, and the second is the state that is being changed. As the Universe moves and changes its state in space in every moment, so there is only one thing that is persistent in that Universe, and that is the occurrence of change. For an engineer, a system that changes its state (points of interests) with a significant rate of change is a dynamical system. This window of time can be a few thousand years for an astrophysicist and a few hundred years for the geologist who is considering the formations of canyons and gorges. Similarly, a financial engineer plans the economy for few years to monthly basis. A mechanical engineer is normally interested in minute scale and a communications engineer worries about milliseconds of delay. This discussion and examples of dynamical systems can go longer and longer as this approach is applicable to each and every subsystem of the Universe at all levels of time scales. But to define an isolated part of reality, it is important to choose the right size time scale and the right set of state variables.

1.1 State Variables and State Vector

State variables are a set of defined variables that are required to completely describe a system for its purpose. These are variables of interests and the minimum set of these variables that completely define the system are assembled together for the description of the system. The combined set of these variables, whether minimum or redundant, is called state vector. For a control systems engineer, the time-dependent state vector is represented as $\vec{x}(t)$ and each independent variable is defined as x followed by a numeral $x_1(t)$, $x_2(t)$, etc.

1.2 Input and Output

Each system interacts with its surroundings, universe, or its environment through the input or the output. Whatever is getting in by a system from the environment is the input and whatever is passing out to the environment is the output. Input describes the changes of a system posed from its surroundings whether these are big, small, desirable, or inappropriate. Whatever a system is contributing to its surroundings, whether asked for or not, is the output of this system and usually an input to some other system. A car takes air from its environment for combustion and this is an input to the car and it runs on the road at a desirable speed. The desirable speed is an output of the car, along with the gases passed from its exhaust. Those gases are also the input to the ecological system of the city providing air to the car. The desired input that is also a function of time is usually denoted by $u(t)$ and a vector $\vec{u}(t)$ for multiple inputs. Similarly required outputs or objectives are denoted by a vector $\vec{y}(t)$.

1.3 Modeling

An engineer requires a model of a system to work with. A model is the basic representation of a system such as graphical, conceptual, or analytical, etc. The Oxford English Dictionary has eight definitions of "model" and philosophically all definitions agree that a model is a simplified representation of limited part of reality with related elements. For an engineer this limited part of reality is a system and a construction of visual, conceptual, or mathematical representation is called modeling. So a model becomes a representation developed specifically to understand, design, study, or analyze the system. This can be a graph on a paper, a flow chart, a mathematical equation, or a small-scale replica of a system or process. There are a few key factors in a modeling scheme for the appropriate representation of a system.

Competence: If a model is not adequate to describe the complete interest in the system than this model may not fulfill the required or intended objectives. If all the required states of a system are not considered, or the relationship between inputs to outputs is vaguely defined, then this may result in an incompetent model. For example, if a transport company only models their services with constant fuel prices, then that company may go out of business more quickly than imagined in the current world scenario.

Simplicity: Contrary to competence, a model must be simple; if a simpler model is possible then one must try to represent it accordingly. There may be states of a system that are not required to describe the complete or unnecessary details. For example, if an engineer designing a braking system of a car also considers the gravitational pull from the stars to the car, he thus makes a very complex and unnecessary model.

Causality: The system may be dependent upon different inputs or requirements that must be considered appropriately. A shuttle in space can never rely upon air supply for combustion. Similarly, an engineer cannot provide a water-cooled model for cooling of machinery in a desert. All dependencies must be considered properly, including time-dependent scenarios such as linking of events, as one cannot occur until the other starts.

Redundancy: This refers to variables or states that can be avoided because some other variable is doing the similar and required work. But reliable systems have some redundancy because if one fails to deliver, the other may help to achieve the objective. It is the same as driving a car in a city where fuel is readily available or driving on a long route highway where fuel may or may not be available easily. A cautious approach is to fill the tank whenever you get a chance for uncertain travel conditions. Redundant systems are generally more reliable but they require more cost to operate or consume more energy. An engineer should not always avoid redundancies but tradeoff between cost and reliability for the better solution. Redundancy also exists in natural systems where required, such as two eyes, two ears, two nostril paths, two lungs, but one tongue.

Experimental Validation: If a model behaves like a real system as conceived, designed, and analyzed through experimental trails then this model is more suitable than any other model that is untested. But usually experimental validation is not possible for all systems and is often a costly requirement for developing prototypes followed by the actual product delivery. A car model can be tested on roads rigorously before delivery to customers, but in the fashion design industry a model may wear new trendy clothes and catwalk.

1.4 Analytical Modeling

The first question that arises is why do analytical or mathematical modeling? Obviously for some people mathematics is a tough subject and they may be happier with only graphical representations or artistic models of the system under study. There are a few famous quotes about mathematics which are very useful in understanding this question.

- There is no branch of mathematics, however abstract, which may not some day be applied to phenomena of the real-world (*Nikolai Lobachevsky*)
- This, therefore is mathematics: she reminds you of the invisible of the soul; she gives life to her own discoveries; she awakens the mind and purifies the intellect; she brings light to our intrinsic ideas; and she abolishes oblivion and ignorance which are ours by birth (*Proclus*)

- From the intrinsic evidence of this creation, the Great Architect of the Universe now begins to appear as a pure mathematician (*Sir James Jeans*)
- This universe is created by an Entity who knows all the Mathematics behind it. Who knows how to solve, when, and where to solve the right variable at right time by the right method (*Author of this book—Asif Mahmood Mughal*)

Mathematical modeling plays a thorough and pivotal rule in understanding and exploring natural phenomenon. A Navier-Stroke's equation can be applied to determine where the wind will take a falling leaf or understand the flow of blood in the human arteries. Tidal waves in oceans, light rays from sun, northern lights (aurora) or power transmission through different medium, all can be explained with electromagnetic and wave equations. A golf ball shot straight to the hole, an off swing delivery to clean bold, a hawk's dive to catch a fish, an airplane flying to its destination are a few examples of balancing aerodynamics forces. Pade' approximation is used to estimate delays in communication channels as well as delays in a feedback loop in the central nervous system of the human body. Mathematical models are applied as financial engineering tools to predict slumps in markets which act as forcing functions for small companies to either quit or merge into large business. The supply and demand of daily needs, upcoming trends in people purchases, and data mining to boost sales are examples of economical modeling of globalization. When a system is modeled in the form of mathematical equations, it will be easy to understand, interpret, or change accordingly for a variety of perspectives. There are different kinds or classifications of mathematical models.

Lumped System Model: An ordinary differential equation with time as the dependent variable is often known as the lumped system model; it changes its state with time or it may change itself with time. A particular definition of lumped system is a system with only one variable of change (dependant variable) and a finite number of states (number of elements in state vector).

Distributed Model: A model with two or more independent variables of change is a distributed model. It is a system that may change with time as well as with spatial coordinates. A good example of this system is weather, which is always changing with time and location. A distributed model occurs with more than one independent variable and the partial differential equation is required to represent this system. A system with infinite number of states is also known as distributed system, e.g., a sample time delay in a system which may have infinite real numbers to fill between two consecutive samples.

Autonomous System Model: A dependent variable does not appear explicitly in the modeling equation and only appears in solutions. So variables can be defined as $x(t) = x$ and in general it is time invariant system. Most time invariant models are autonomous such as the application of Newton's second law of motion on ordinary objects, $F = m \cdot \ddot{x}(t) = m \cdot a$. Where acceleration and applied force are functions of time, but variable of time t does not appear when values or equations for force or acceleration are quantified.

Time Varying or Time Invariant System: A model that changes with time is a time varying model; it means the mathematical equation that is representing the system also changes with time. So for a one specific instant a modeling equation is applicable but after some time another modeling equation is governing the same system. Large air carriers, space shuttles, and other high fuel consuming objects are considered time varying because of their rapid change of mass during their journey. Newton's second law of motion is represented as $F(t) = m(t) \cdot \ddot{x}(t)$ for these time varying systems. But if we treat $m(t) = m$, constant mass during its operation then this is a time invariant system, though force and acceleration are changing with time. A system which is considered unchanged in physical properties for some specific time is known as the time invariant system.

Linear and Nonlinear Systems: If an input excites a system to perform some actions under specific conditions and this can produce the same results for the same input and conditions in every other time, then it is a linear system. In simple words, the principle of superposition should apply to the system. Let us consider a system $y(t) = x(t) + u(t)$ where two different results from two different values and inputs are given as $y_1(t) = x_1(t) + u_1(t)$ and $y_2(t) = x_2(t) + u_2(t)$ and if their homogenous and additive combination can be obtained for $x_3(t) = \alpha \cdot x_1(t) + \beta \cdot x_2(t)$ and $u_3(t) = \alpha \cdot u_1(t) + \beta \cdot u_2(t)$ as $y_3(t) = \alpha \cdot y_1(t) + \beta \cdot y_2(t)$, then the system is linear. If this principle cannot be applied to a system then that system is nonlinear. An example is if either of $x(t)$ or $u(t)$ are trigonometric functions or other algebraic functions where superposition is not applicable.

Continuous or Discrete Systems: If a system is defined at each and every instant of time, so that its solution can be obtained for each instant of time from the smallest fraction of a microsecond to several thousands of years, then it is a continuous system. If in order to define a system or to obtain its solution, we have to sample the time instants according to our convenience, then this is a discrete system. A discrete system is silent between two intervals and choosing the sampling interval as smallest as possible makes it closest to a continuous system. A car runs on road continuously but a microcontroller-based electronic fuel injection system of a car processes the combustion in discrete samples.

Causal and Non-causal System: A system that only depends upon current and previous values to give output is a causal system. A causal system holds a memory for some past events in accumulation with current scenario to produce results. Causal systems do not depend upon the future or predict or anticipate anything for output, unlike a non-causal system that does depend upon future inputs to have current output results. All physical systems are causal in nature such as mass springs, dampers, capacitors, and inductors which hold some previous stretches or values in addition to applied forces to give present output. Non-causal systems can only exist in virtual worlds and are used to solve computer algorithms or estimation problems.

1.5 Physical and Computational Systems

All physical systems in the Universe are causal, continuous, nonlinear, and time varying (and distributed) in reality. So why study Acausal (non-causal), discrete, linear, and time-invariant lumped systems? The only answer is ease to handle the problem. The laws of nature do not require future inputs to solve for current problems, as causal systems use only previous history and current inputs to solve for present output and future responses with some valid assumptions. But we require developing an algorithm which predicts future states and satisfy the current requirements, so a computer program may be non-causal. If we like to solve an exact solution of a small problem, it is possible to do so but for large problems continuous systems require memory storage and computation speed that may not be available to us. Thus it requires to discretize the system with such close sampling interval that it seems to behave like a continuous system. Any continuous system problem solver in computer is actually a discrete problem solver with very small sampling time. The principal of superposition does not hold in nature in an exact sense; whatever we get today may not be equal to what we get tomorrow in actuality. Somewhere some depreciation or addition has been done. But we study linear systems because superposition principal makes us to solve analytical models with great ease. We can easily assume that under certain conditions the solution of linear and nonlinear will be so close that their relative error is meaningless to us. Similarly a dynamical system is the one that changes with time but to monitor change in each and every parameter, make the model with almost infinite states very difficult to solve analytically. Anything that comes on the axis of time is bound to change but monitoring change of a state is relatively very easy as to monitor change in physical properties. A book may be losing its shine or binding but if never read or issued is in a constant condition for a librarian. Changing values of resistors on a minute scale due to aging and environment may be useless when designing a circuit. A car's motion on the road is represented by the time invariant system $F = m \cdot a$ where m is constant for the car, though there is continuous consumption of fuel and neglecting variation of total mass by assuming $m(t) \approx m$ for maximum fuel and loading capacity. This assumption is not valid for a rocket taking a shuttle to the space.

1.6 Mathematical Model

A simple required mathematical model is a set of differential equations governing the dynamics of a system. The sum of the highest derivatives of all independent variables becomes the order of the system. A modeling equation that does not have any derivative becomes a simple algebraic equation. A third-order ordinary differential equation (ODE) in Eq. (1.1) represents a model of a physical system.

$$\dddot{z}(t) + a(t) \cdot \ddot{z}(t) + b(t) \cdot \dot{z}(t) + c(t) \cdot z(t) + d(t) \cdot \delta(t) = 0 \tag{1.1}$$

In this equation $a(t)$, $b(t)$, $c(t)$ and $d(t)$ are time varying physical parameters, and it is a linear time varying model of a system. If we consider these parameters as constant a, b, c and d then this becomes a linear time invariant (LTI) system. This is a causal system and requires initial conditions to solve for its solution. It is difficult to solve higher-order different ODEs and solution of first-order ODE is simple. It is better and convenient to solve three ODEs than to solve a single third-order ODE. The Eq. (1.1) can be represented as three first-order ODEs. Let $x_1(t) = z(t)$, $x_2(t) = \dot{z}(t) = \dot{x}_1(t)$, and $x_3(t) = \ddot{z}(t) = \dot{x}_2(t)$ and Eq. (1.1) is now represented in matrix form for the three first-order ODEs

$$\begin{bmatrix} \dot{x}_1(t) \\ \dot{x}_2(t) \\ \dot{x}_3(t) \end{bmatrix} = \begin{bmatrix} 0 & 1 & 0 \\ 0 & 0 & 1 \\ -c & -b & -a \end{bmatrix} \cdot \begin{bmatrix} x_1(t) \\ x_2(t) \\ x_3(t) \end{bmatrix} + \begin{bmatrix} 0 \\ 0 \\ -d \end{bmatrix} \cdot \delta(t) \tag{1.2}$$

The third equation in Eq. (1.2) is same as Eq. (1.1) with a change of variable from $z(t)$ to $x_3(t)$, but in order to solve for first order $x_3(t)$, it is required to solve $x_1(t)$ and $x_2(t)$. The solution of a single first-order differential equation and three first-order equations simultaneously is the same algebraically. Methods of linear algebra provide solution to a matrix differential equation (1.2).

1.7 State Space Method

State Space equations are a set of first-order differential equations in a specific matrix notation. The state vector is a set of state variables required to model the system. An input vector and an output vector constitute the set of all inputs and outputs respectively in a vector form. Equation (1.1) is written in state space form as Eq. (1.2) if we define a state vector and input vector as

$$\vec{x}(t) = \begin{bmatrix} x_1(t) \\ x_2(t) \\ x_3(t) \end{bmatrix}, \quad \vec{u}(t) = [\delta(t)] \tag{1.3}$$

Let us define two outputs as $\dot{z}(t) + k \cdot \delta(t)$ and $e(t) \cdot z(t) + h(t) \cdot \ddot{z}(t)$. So two outputs are in a vector form and with definition of state variables are given as

$$\begin{aligned} \vec{y}(t) &= \begin{bmatrix} y_1(t) \\ y_2(t) \end{bmatrix} = \begin{bmatrix} x_2(t) + k \cdot u(t) \\ e(t) \cdot x_1(t) + h(t) \cdot x_3(t) \end{bmatrix} \\ &= \begin{bmatrix} 0 & 1 & 0 \\ e(t) & 0 & h(t) \end{bmatrix} \cdot \begin{bmatrix} x_1(t) \\ x_2(t) \\ x_3(t) \end{bmatrix} + \begin{bmatrix} k \\ 0 \end{bmatrix} \cdot u(t) \end{aligned} \tag{1.4}$$

The Eqs. (1.2) and (1.4) represent the model from input to output as state space representation using state, input, and output vectors in Eqs. (1.3) and (1.4). These equations can be represented generally for any linear system as

$$\begin{aligned} \dot{\vec{x}}(t) &= A(t) \cdot \vec{x}(t) + B(t) \cdot \vec{u}(t) \\ \vec{y}(t) &= C(t) \cdot \vec{x}(t) + D(t) \cdot \vec{u}(t) \end{aligned} \tag{1.5}$$

State space representation for a nonlinear system is given as

$$\begin{aligned} \dot{\vec{x}}(t) &= f\left(\vec{x}(t), \vec{u}(t), t\right) \\ \vec{y}(t) &= g\left(\vec{x}(t), \vec{u}(t), t\right) \end{aligned} \tag{1.6}$$

Both linear and nonlinear systems have first-order derivatives for all state variables. If state and input vectors can be separated through superposition then the system is linear as in Eq. (1.5), otherwise they are nonlinear. The time variance depends upon the existence of variable t explicitly and state equation and output equation in Eqs. (1.5) and (1.6) are functions of time. The system is LTI if matrices in Eq. (1.5) are constant.

$$\begin{aligned} &A(t) = A \quad B(t) = B \quad C(t) = C \quad D(t) = D \\ &\dot{\vec{x}}(t) = A \cdot \vec{x}(t) + B \cdot \vec{u}(t) \\ &\vec{y}(t) = C \cdot \vec{x}(t) + D \cdot \vec{u}(t) \end{aligned} \tag{1.7}$$

For nonlinear time invariant system Eq. (1.6) is modified as

$$\begin{aligned} \dot{\vec{x}}(t) &= \vec{f}\left(\vec{x}(t), \vec{u}(t)\right) \\ \vec{y}(t) &= \vec{g}\left(\vec{x}(t), \vec{u}(t)\right) \end{aligned} \tag{1.8}$$

For LTI system the order of matrices A, B, C, D depends upon the number of states, inputs, and outputs. The state vector $\vec{x}(t)$ is $n \times 1$ for n state variables, $\vec{u}(t)$ is a $p \times 1$ vector for p inputs, and $\vec{y}(t)$ is $q \times 1$ vector for q outputs of the system. This definition of these vectors determines the order of matrices as:

$$A \in F^{n \times n} \quad B \in F^{n \times p} \quad C \in F^{q \times n} \quad D \in F^{q \times p}$$

The matrix A is always square and the order of matrix D represents the direct input–output relation. In Eq. (1.5) A is 3×3 matrix, and only one input so $\vec{u}(t)$ is 1×1 and so B is 3×1. For two outputs, $\vec{y}(t)$ is 2×1 vector and C is 2×3 matrix and finally for two outputs and single input D is 2×1. For nonlinear systems there are only $n \times 1$ state equations. The vector field F is either real space $\mathbb{R}$ or complex space $\mathbb{C}$ depending upon the system; and for all physical systems, these matrices belong to real space $\mathbb{R}$. These matrices may belong to complex space $\mathbb{C}$ for a virtual system or computational purposes, but cannot be physically implemented through real devices.

1.8 Linearization

The nonlinear state space system is linearized to obtain linear state space equations. The mathematical principle behind the concept is Taylor's series approximation for defining any function within a specified range. Let $\vec{x}_e$ and $\vec{u}_e$ be equilibrium state and input at equilibrium point respectively and let Δx, Δu be the small perturbations from the equilibrium where the system is linear. The nonlinear functions in $n \times 1$ vector for state space with n state variables and p input variables is given as

$$\dot{\vec{x}}(t) = \vec{f}\left(\vec{x}(t), \vec{u}(t)\right) \tag{1.9}$$

At equilibrium point the function is relaxed

$$\vec{f}\left(\vec{x}_e, \vec{u}_e\right) = 0 \tag{1.10}$$

We can write expression by expanding as follows

$$\dot{\vec{x}}(t) = \begin{bmatrix} \dot{x}_1 \\ \dot{x}_2 \\ \vdots \\ \dot{x}_n \end{bmatrix} = \begin{bmatrix} f_1(x_1, x_2, \cdots x_n, u_1, u_2 \cdots u_p) \\ f_2(x_1, x_2, \cdots x_n, u_1, u_2 \cdots u_p) \\ \vdots \\ f_n(x_1, x_2, \cdots x_n, u_1, u_2 \cdots u_p) \end{bmatrix} \tag{1.11}$$

As we perturb the state $\vec{x}$ within small region $\Delta\vec{x}$ from the equilibrium $\vec{x}_e$ at given inputs $\vec{u}_e$ such as

$$\vec{x}(t) = \vec{x}_e + \Delta\vec{x}(t) \tag{1.12}$$

$$\vec{u}(t) = \vec{u}_e + \Delta\vec{u}(t) \tag{1.13}$$

So Eq. (1.9) is given as

$$\dot{\vec{x}}(t) = \vec{f}\left(\vec{x}_e + \Delta\vec{x}(t), \vec{u}_e + \Delta\vec{u}(t)\right) \tag{1.14}$$

The Taylor series expansion of vector functions is given as

$$\dot{\vec{x}}(t) = \vec{f}\left(\vec{x}_e + \Delta\vec{x}(t), \vec{u}_e + \Delta\vec{u}(t)\right) \tag{1.15}$$

$$\frac{d\vec{x}(t)}{dt} \cong \vec{f}\left(\vec{x_e}, \vec{u}_e\right) + \frac{\partial \vec{f}}{\partial \vec{x}}\left(\vec{x_e}, \vec{u}_e\right) \cdot \Delta\vec{x} + \frac{\partial \vec{f}}{\partial \vec{u}}\left(\vec{x_e}, \vec{u}_e\right) \cdot \Delta\vec{u}$$
$$+ \text{higher order terms} \tag{1.16}$$

We know that $\vec{f}\left(\vec{x_e}, \vec{u}_e\right) = 0$ and we can ignore second-order and higher-order terms by assuming negligible effect within the perturbation region $\Delta\vec{x}(t)$ and $\Delta\vec{u}(t)$. We also observe that the equilibrium point is a constant so

$$\dot{\vec{x}}(t) = \frac{d\vec{x}(t)}{dt} = \frac{d\left(\vec{x}_e + \Delta\vec{x}(t)\right)}{dt} = \frac{d\left(\Delta\vec{x}(t)\right)}{dt} = \frac{d\Delta\vec{x}}{dt} = \Delta\dot{\vec{x}}. \tag{1.17}$$

Now a linear state space is given as

$$\Delta\dot{\vec{x}} = \frac{\partial \vec{f}}{\partial \vec{x}}\left(\vec{x_e}, \vec{u}_e\right) \cdot \Delta\vec{x} + \frac{\partial \vec{f}}{\partial \vec{u}}\left(\vec{x_e}, \vec{u}_e\right) \cdot \Delta\vec{u} = A \cdot \Delta\vec{x} + B \cdot \Delta\vec{u} \tag{1.18}$$

Where state matrix A and input gain matrix B are computed Jacobian matrices at equilibrium points

$$A = \frac{\partial \vec{f}}{\partial \vec{x}}\left(\vec{x_e}, \vec{u}_e\right) = \left.\begin{bmatrix} \frac{\partial f_1}{\partial x_1} & \frac{\partial f_1}{\partial x_2} & \cdots & \frac{\partial f_1}{\partial x_n} \\ \frac{\partial f_2}{\partial x_1} & \frac{\partial f_2}{\partial x_2} & \cdots & \frac{\partial f_2}{\partial x_n} \\ \vdots & \vdots & \ddots & \vdots \\ \frac{\partial f_n}{\partial x_1} & \frac{\partial f_n}{\partial x_2} & \cdots & \frac{\partial f_n}{\partial x_n} \end{bmatrix}\right|_{\substack{\vec{x} = \vec{x_e} \\ \vec{u} = \vec{u_e}}} \tag{1.19}$$

$$B = \frac{\partial \vec{f}}{\partial \vec{u}}\left(\vec{x_e}, \vec{u}_e\right) = \left.\begin{bmatrix} \frac{\partial f_1}{\partial u_1} & \frac{\partial f_1}{\partial u_2} & \cdots & \frac{\partial f_1}{\partial u_p} \\ \frac{\partial f_2}{\partial u_1} & \frac{\partial f_2}{\partial u_2} & \cdots & \frac{\partial f_2}{\partial u_p} \\ \vdots & \vdots & \ddots & \vdots \\ \frac{\partial f_n}{\partial u_1} & \frac{\partial f_n}{\partial u_2} & \cdots & \frac{\partial f_n}{\partial u_p} \end{bmatrix}\right|_{\substack{\vec{x} = \vec{x_e} \\ \vec{u} = \vec{u_e}}} \tag{1.20}$$

The outputs nonlinear equation is given as follows

$$\vec{y}(t) = \vec{g}\left(\vec{x}(t), \vec{u}(t)\right) = \vec{g}\left(\vec{x}_e + \Delta\vec{x}(t), \vec{u}_e + \Delta\vec{u}(t)\right) \tag{1.21}$$

It can also be linearized at the same equilibrium point using Taylor series expansion in a similar way. The Output-state matrix C and output–input gain matrix D are also computed as Jacobian matrices as follows

$$C = \frac{\partial \vec{y}}{\partial \vec{x}}\left(\vec{x_e}, \vec{u}_e\right) = \left.\begin{bmatrix} \frac{\partial y_1}{\partial x_1} & \frac{\partial y_1}{\partial x_2} & \cdots & \frac{\partial y_1}{\partial x_n} \\ \frac{\partial y_2}{\partial x_1} & \frac{\partial y_2}{\partial x_2} & \cdots & \frac{\partial y_2}{\partial x_n} \\ \vdots & \vdots & \ddots & \vdots \\ \frac{\partial y_q}{\partial x_1} & \frac{\partial y_q}{\partial x_2} & \cdots & \frac{\partial y_q}{\partial x_n} \end{bmatrix}\right|_{\substack{\vec{x} = \vec{x_e} \\ \vec{u} = \vec{u_e}}} \tag{1.22}$$

$$D = \frac{\partial \vec{y}}{\partial \vec{x}}\left(\vec{x_e}, \vec{u}_e\right) = \left.\begin{bmatrix} \frac{\partial y_1}{\partial u_1} & \frac{\partial y_1}{\partial u_2} & \cdots & \frac{\partial y_1}{\partial u_p} \\ \frac{\partial y_2}{\partial u_1} & \frac{\partial y_2}{\partial u_2} & \cdots & \frac{\partial y_2}{\partial u_p} \\ \vdots & \vdots & \ddots & \vdots \\ \frac{\partial y_q}{\partial u_1} & \frac{\partial y_q}{\partial u_2} & \cdots & \frac{\partial y_q}{\partial u_p} \end{bmatrix}\right|_{\substack{\vec{x} = \vec{x_e} \\ \vec{u} = \vec{u_e}}} \tag{1.23}$$

The orders of the computed Jacobian matrices are same as

$$A \in \mathbb{R}^{n \times n} \quad B \in \mathbb{R}^{n \times p} \quad C \in \mathbb{R}^{q \times n} \quad D \in \mathbb{R}^{q \times p}$$

1.9 Transfer Function

The relation between outputs to input is called transfer function and mathematically it is a rational function $G(s)$ of a single variable. This single variable expresses the mutual relationship of different states and also the relationship from input to states and then to states to output. Transfer functions are representatives of a system in frequency domain in the form of algebraic equations. The mathematical relationship in frequency domain between states $X(s)$, input $U(s)$, and output $Y(s)$ is given as

$$G(s) = \frac{Y(s)}{X(s)} \cdot \frac{X(s)}{U(s)} = \frac{Y(s)}{U(s)} \tag{1.24}$$

There is a single transfer function for single input to single output (SISO) system; but for multiple inputs multiple outputs (MIMO) system, transfer function $G(s)$ is a matrix of $q \times p$ order. Each entry in the matrix transfer function represents the relationship of a specific output to a specific input. There is a relationship between transfer function and state space matrices, which is known as realization of a system in frequency or time domain. State space is realization in time domain and taking the Laplace transform of Eq. (1.5) for zero initial conditions provides realization in frequency domain as $G(s)$. A system with zero initial condition is known as a *relaxed system*. Taking Laplace transform of Eq. (1.5) as LTI yields

$$\begin{aligned} sX(s) - x(t_0) &= A \cdot X(s) + B \cdot U(s) \\ Y(s) &= C \cdot X(s) + D \cdot U(s) \end{aligned} \tag{1.25}$$

For relaxed system $x(t_0) = x(0) = 0$ and so transfer function (TF) $G(s)$ is

$$\vec{G}(s) = C \cdot [s \cdot I - A]^{-1} \cdot B + D \tag{1.26}$$

For the system in Eqs. (1.2) and (1.4), there are two outputs and an input, so there are two transfer functions in a matrix. The first and second rows of C matrix correspond to $Y_1(s)$ and $Y_2(s)$ respectively.

$$\vec{G}(s) = \begin{bmatrix} \frac{Y_1(s)}{U(s)} \\ \frac{Y_2(s)}{U(s)} \end{bmatrix} = \begin{bmatrix} 0 & 1 & 0 \\ e(t) & 0 & h(t) \end{bmatrix} \cdot \begin{bmatrix} s & -1 & 0 \\ 0 & s & -1 \\ c & b & a \end{bmatrix}^{-1} \cdot \begin{bmatrix} 0 \\ 0 \\ -d \end{bmatrix} + \begin{bmatrix} k \\ 0 \end{bmatrix} \tag{1.27}$$

The typical definition of a TF comes as a rational function with a numerator and denominator as algebraic equation.

$$G(s) = \frac{N(s)}{D(s)} = \frac{\beta_m \cdot s^m + \beta_{m-1} \cdot s^{m-1} \ldots\ldots \beta_2 \cdot s^2 + \beta_1 \cdot s + \beta_0}{s + \alpha_{n-1} \cdot s^{n-1} \ldots\ldots \alpha_2 \cdot s^2 + \alpha_1 \cdot s + \alpha_0} \tag{1.28}$$

The order of numerator m is always less or equal to the order of denominator n for a proper rational function. If $m = n$ then the corresponding entry in D is non zero and if $m < n$ then corresponding entry in D matrix is 0. In Eq. (1.27), the numerator of $\frac{Y_1(s)}{U(s)}$ has order equal to denominator which is 3, but for $\frac{Y_2(s)}{U(s)}$ the order of numerator should be less than 3. The order of denominator is always equal to the order of square matrix A and determinant of $[s \cdot I - A]$ is denominator for all transfer functions in $\vec{G}(s)$.

Example 1.1: State Space Equations Formulate a state space representation for the following set of ODE and obtain transfer function matrix.

$$\begin{aligned} \ddot{\theta} + 2\dot{\theta} + 3\theta - 2x &= \tau + \frac{F}{2} \\ \dot{x} - 5\theta &= F - \frac{\tau}{5} \end{aligned} \tag{1.29}$$

Solution The set of state variables consists of $\theta, \dot{\theta}$, and x. The total order of the system is three by summing highest derivative terms in all variables. Let us define a state and input vectors $\vec{x}$ and $\vec{u}$ as

$$\begin{array}{ll} x_1 = \theta & \\ x_2 = \dot{\theta} = \dot{x}_1 & u_1 = \tau \\ x_3 = x & u_2 = F \end{array} \tag{1.30}$$

So Eq. (1.29) can be written as

$$\begin{aligned} \dot{x}_2 &= -3x_1 - 2x_2 + 2x_3 + u_1 + \frac{u_2}{2} \\ \dot{x}_3 &= 5x_1 - \frac{u_1}{5} + u_2 \end{aligned} \tag{1.31}$$

So the state space representation is

$$\begin{bmatrix} \dot{x}_1 \\ \dot{x}_2 \\ \dot{x}_3 \end{bmatrix} = \begin{bmatrix} 0 & 1 & 0 \\ -3 & -2 & 2 \\ 5 & 0 & 0 \end{bmatrix} \cdot \begin{bmatrix} \dot{x}_1 \\ \dot{x}_2 \\ \dot{x}_3 \end{bmatrix} + \begin{bmatrix} 0 & 0 \\ 1 & \frac{1}{2} \\ -\frac{1}{5} & 1 \end{bmatrix} \cdot \begin{bmatrix} u_1 \\ u_2 \end{bmatrix} \tag{1.32}$$

This is state space form given as in Eq. (1.7) as LTI system with $A = \begin{bmatrix} 0 & 1 & 0 \\ -3 & -2 & 2 \\ 5 & 0 & 0 \end{bmatrix}$ and $B = \begin{bmatrix} 0 & 0 \\ 1 & \frac{1}{2} \\ -\frac{1}{5} & 1 \end{bmatrix}$. The output in this equation is not explicitly mentioned and so in this case, we can measure all the state variables at output. The output equation will be

$$\vec{y}(t) = I_{3\times 3} \cdot \vec{x}(t) \tag{1.33}$$

In this case, $C = \begin{bmatrix} 1 & 0 & 0 \\ 0 & 1 & 0 \\ 0 & 0 & 1 \end{bmatrix}$ and $D = \begin{bmatrix} 0 & 0 \\ 0 & 0 \\ 0 & 0 \end{bmatrix}$ in output equation. Matrix D is often a null matrix and null matrices are not written in state space representations. MATLAB command **ss2tf** gets a TF from state space representation and vice versa for **tf2ss** command. Using symbolic math, the transfer function can directly be computed as

```
syms s
A=[0 1 0;-2 -2 2;5 0 0]
B=[0 0;1 1/2;-1/5 1]
C=eye(3,3)
G=C*inv(s*eye(3,3)-A)*B
simplify(G)
```

This generates $2 \times 3\ \vec{G}(s)$ matrix of transfer function for three outputs and two inputs. All denominators in TF are same, whereas there are six different numerators

$$\vec{G}(s) = \frac{1}{s^3 + 2s^2 + 2s - 10} \cdot \begin{bmatrix} s - 0.4 & 0.5s + 2 \\ s^2 - 0.4 & 0.5s^2 + 2 \\ -0.2s^2 - 0.4s + 4.6 & s^2 + 2s + 4.5 \end{bmatrix} \quad (1.34)$$

Example 1.2: State Space and Transfer Function Formulate a state space representation and transfer function matrix of

$$\dddot{z}(t) - 2\ddot{z}(t) - 4\dot{z}(t) + 8z(t) = f_1(t) + 8f_2(t), \quad (1.35)$$

for output $m(t) = 3z(t) + 5f_1(t)$

Solution The state, input, and output vectors are

$$\begin{matrix} x_1(t) = z(t) \\ x_2(t) = \dot{x}_1(t) = \dot{z}(t) \\ x_3(t) = \dot{x}_2(t) = \ddot{z}(t) \end{matrix} \quad \begin{matrix} u_1(t) = f_1(t) \\ u_2(t) = f_2(t) \end{matrix} \quad y(t) = m(t) \quad (1.36)$$

The state space representation for LTI system is

$$\begin{aligned} \dot{\vec{x}}(t) &= \begin{bmatrix} 0 & 1 & 0 \\ 0 & 0 & 1 \\ -8 & 4 & 2 \end{bmatrix} \cdot \vec{x}(t) + \begin{bmatrix} 0 & 0 \\ 0 & 0 \\ 1 & 8 \end{bmatrix} \cdot \vec{u}(t) \\ \vec{y}(t) &= \begin{bmatrix} 3 & 0 & 0 \end{bmatrix} \cdot \vec{x}(t) + \begin{bmatrix} 5 & 0 \end{bmatrix} \end{aligned} \quad (1.37)$$

$$\vec{G}(s) = \frac{1}{s^3 - 2s^2 - 4s + 8} \cdot \begin{bmatrix} 5s^3 - 10s^2 - 20s + 43, & 24 \end{bmatrix} \quad (1.38)$$

Note: If there is nonzero term in D then the corresponding numerator has the same order as denominator

Example 1.3: Nonlinear State Space Equations Formulate a state space representation for the following nonlinear system

$$\begin{aligned} \ddot{y}(t) + 3(\dot{y}(t))^2 - 3\cos(y(t)) + \sin(u(t)) &= 1 \\ \dot{y}(t) - 2 \cdot cos(y(t)) &= -2z(t) \\ \dot{z}(t) + \sqrt{(z(t))} &= 2\sin(y(t)) \\ \ddot{z}(t) &= u(t) - p(t) \end{aligned} \quad (1.39)$$

where output is $\eta(t) = (\dot{y}(t))^2 + sin(p(t)) - \cos(z(t))$. Later find the transfer function of the system by linearizing at $\vec{x_e} = [\ \pi/4 \quad 1 \quad 4 \quad 0\ ]^T$ and $\vec{u_e} = [0 \quad 0]^T$.

Solution The state and input vector is

$$\begin{array}{ll} x_1(t) = y(t) & \\ x_2(t) = \dot{x}_1(t) = \dot{y}(t) & u_1(t) = u(t) \\ x_3(t) = z(t) & u_2(t) = u(t) \\ x_4(t) = \dot{x}_3(t) = \dot{z}(t) & \end{array} \tag{1.40}$$

The state space representation for four state variables and two inputs is given as

$$\begin{aligned} \dot{x}_1(t) &= f_1\left(\vec{x}(t), \vec{u}(t), t\right) &&= 2 \cdot cos(x_1(t)) - 2x_3(t) \\ \dot{x}_2(t) &= f_2\left(\vec{x}(t), \vec{u}(t), t\right) &&= 3\cos(x_1(t)) - 3(x_2(t))^2 + \sin(u_1(t)) + 1 \\ \dot{x}_3(t) &= f_3\left(\vec{x}(t), \vec{u}(t), t\right) &&= 2\sin x_1(t) - \sqrt{(x_3(t))} \\ \dot{x}_4(t) &= f_4\left(\vec{x}(t), \vec{u}(t), t\right) &&= u_1(t) - u_2(t) \end{aligned} \tag{1.41}$$

The single output equation is given as

$$\vec{y}(t) = \eta(t) = g_1\left(\vec{x}(t), \vec{u}(t), t\right) = (x_2(t))^2 - \cos(x_3(t)) + sin(u_2(t)) \tag{1.42}$$

Linearization of the system is a cumbersome process but can easily be done by using the commands of the symbolic toolbox of MATLAB. A few commands are shown below to demonstrate the process

```
syms x1 x2 x3 x4 u1 u2
f1=2*cos(x1)-2*x3
f2=3*cos(x1)-3*x2*x2+sin(u1)+1
f3=2*sin(x1)-sqrt(x3)
f4=u1-u2
a11=diff(f1,x1)
b12=diff(f1,u2)
c13=diff(g1,x3)
C=[c11 c12 c13 c14]
D=[d11 d12]
x1=pi/4; x2=1; x3=4; x4=0; u1=0; u2=0;
eval(C)
```

Transfer function can be found accordingly from state space realization as

$$G(s) = \begin{bmatrix} \dfrac{2}{s+6} & 1 \end{bmatrix} \tag{1.43}$$

Problems

P1.1 **Input–output**
Explain the input–output of different subsystems of a car, how is the input of one system derived from the output of the other system?

P1.2 **Redundant Systems**
Redundancies increase reliability of the system but also increases costs. Are expensive systems always reliable? Explain your answer with examples, models, and system.

P1.3 **State Space of Linear Systems**
Solve to find state space representation and TF of the following systems

(a)
$$2 \cdot \ddot{y}(t) - 4 \cdot \dot{y}(t) + 6 \cdot y(t) + 2 \cdot \dot{\varphi}(t) = 8 \cdot \delta(t) - \rho(t)$$
$$\ddot{\varphi}(t) - 3 \cdot y(t) = \delta(t)$$
$$h(t) = 5 \cdot \varphi(t) + y(t) + 2 \cdot \delta(t)$$

(b)
$$\ddot{I}(t) + \frac{R}{L}\dot{I}(t) + \frac{1}{LC}I(t) = E(t)$$

(c)
$$\ddot{y}(t) + \omega^2 \cdot y(t) = F(t)$$
$$l(t) = y(t)$$

P1.4 **Linearization for State Space Representation**
Find state space representation of nonlinear system and linearize at relaxed conditions

(a) $l \cdot \ddot{\theta}(t) + g \cdot \sin\theta = f(t)$

(b) $m \cdot \dot{v}(t) = F_g - b \cdot v^2$ and output is $h(t)$ where $v(t) = \dot{h}(t)$

(c)
$$\dot{y}_1(t) = a \cdot y_1(t) - b \cdot y_1(t) \cdot y_2(t)$$
$$\dot{y}_2(t) = k \cdot y_1(t) \cdot y_2(t) - l \cdot y_2(t)$$

References

1. Friedland, Bernard. 2005. *Control System Design—An Introduction to State Space Methods*. Mineola: Dover Publications.
2. Chen, Chi-Tsonga. 1999. *Linear System Theory and Design*, 3rd ed. Oxford: Oxford University Press.
3. Heidi, Christian, André Ran, and Freak van Schrage. 2007. *Introduction to Mathematical Systems Theory—Linear Systems, Identification and Control*. Basel: Birkhäuser Verlag.
4. Rugh, Wilson J. 1996. *Linear System Theory*. Upper Saddle River: Prentice Hall.
5. Franklin, Gene F., J. David Powell, and Michael L. Workman. 1997. *Digital Control of Dynamic Systems*, 3rd ed. Richmond: Addison-Wesley-Longman.
6. Gershenfeld, Neil. 2011. *The Nature of Mathematical Modeling*. New York: Cambridge University Press.
7. Ogata, Katisuhiko. 1995. *Discrete—Time Control Systems*, 2nd ed. Upper Saddle River: Prentice Hall.
8. Hespanha, Joao. 2009. *Linear Systems Theory*. Princeton University Press: Princeton.

Chapter 2
Lagrangian Modeling

The basic laws of physics are used to model every system whether it is electrical, mechanical, hydraulic, or any other energy domain. In mechanics, Newton's laws of motion provide key concepts to model-related physical phenomenon. The Lagrangian formulation of modeling derives from the basic work–energy principle and Newton's laws of motion. The basic law states that the force acting on a body is directly proportional to its acceleration equated with constant mass.

$$F = ma = m\frac{dv}{dt} = m\frac{d^2x}{dt^2} \tag{2.1}$$

In this equation, force is directly proportional to the derivative of velocity v or double derivative of displacement x with respect to time. If a body moves from point A to point B then the work done by a body is a dot product of force and displaced path integrated from point A to B.

$$W = \int_A^B F \cdot dx \tag{2.2}$$

$$W = \int_A^B m\frac{d^2x}{dt^2} \cdot dx = \int_A^B m\ddot{x} \cdot dx \tag{2.3}$$

We now derive the equation as

$$\ddot{x}dx = \frac{d\dot{x}}{dt}dx = d\dot{x}\,\frac{dx}{dt} = \dot{x}\,d\dot{x} \tag{2.4}$$

A.M. Mughal, *Real Time Modeling, Simulation and Control of Dynamical Systems*,
DOI 10.1007/978-3-319-33906-1_2

So Eq. (2.3) formulates as

$$W = \int_A^B m\dot{x}\, d\dot{x} = m\left(\frac{\dot{x}^2}{2}\right)\Big|_A^B = \frac{m}{2}\left(\dot{x}_B^2 - \dot{x}_A^2\right) \tag{2.5}$$

$$W = K_B - K_A = \Delta K \tag{2.6}$$

It is important to know that here we are discussing motion in a conservative field where work done is independent of path and depends upon the difference between the initial and final value of kinetic energy K. We note that work done is represented as change in K between two points. By the law of conversation of energy and work–energy principle we know that a change in kinetic energy should also change the potential energy U, as the total energy of the system remains constant. So a change in kinetic and change in potential energies should have a net effect of zero as

$$\Delta K + \Delta U = 0 \tag{2.7}$$

So

$$\Delta U = -W = -\int_A^B F \cdot dx \tag{2.8}$$

Using the fundamental theorem of calculus for antiderivatives, we express Eq. (2.8) as

$$F = -\frac{dU}{dx} \tag{2.9}$$

This shows forces required to change the potential energy, whereas the force required to change the kinetic energy is determined from the definition of K.

$$K = \frac{1}{2}mv^2 = \frac{1}{2}m\dot{x}^2 \tag{2.10}$$

So

$$\frac{\partial K}{\partial \dot{x}} = m\dot{x} \tag{2.11}$$

Differentiating Eq. (2.11) gives a force of kinetic energy

$$F = \frac{d}{dt}\left(\frac{\partial K}{\partial \dot{x}}\right) = \frac{d}{dt}(m\dot{x}) = ma \tag{2.12}$$

As sum of forces equal to zero so

$$\frac{d}{dt}\left(\frac{\partial K}{\partial \dot{x}}\right) - \frac{dU}{dx} = 0 \quad (2.13)$$

Now we define a Lagrangian function L as the difference of kinetic and potential energies

$$L = K - U \quad (2.14)$$

We observe that kinetic energy K is a function of velocity $\dot{x}$, and potential energy U is a function of displacement x. Accordingly the Lagrangian function L has two terms as a function of velocity $\dot{x}$ and displacement x independently. As we know that

$$\left(\frac{\partial L}{\partial \dot{x}}\right) = \left(\frac{\partial K}{\partial \dot{x}}\right) \text{ and } \frac{\partial L}{\partial x} = \frac{\partial U}{\partial x} \quad (2.15)$$

Lagrangian function L now represents Eq. (2.13) as

$$\frac{d}{dt}\left(\frac{\partial L}{\partial \dot{x}}\right) - \frac{\partial L}{\partial x} = 0 \quad (2.16)$$

This is Lagrangian formulation for a system with motion only in single axis.

2.1 Modeling in Three Axes

Now consider a motion in three axes and every equation now to be represented in three dimensional vector notations.

$$\vec{F} = m\frac{d^2\vec{x}}{dt^2} \quad (2.17)$$

The force $\vec{F}$ has three components, F_x, F_y, and F_z in x, y and z axes respectively. The potential energy U is a scalar field in three dimensions and the negative gradient of this field provides a force vector in three axes. Now Eq. (2.9) represents as

$$\vec{F} = -\nabla\, U(x, y, z) = -\left[\frac{\partial U}{\partial x} \quad \frac{\partial U}{\partial y} \quad \frac{\partial U}{\partial z}\right]^T \quad (2.18)$$

The kinetic energy K in three axes is given as

$$K = \frac{1}{2}m\left\|\vec{\dot{x}}\right\|^2 = \frac{1}{2}m\left(\vec{\dot{x}}^T \cdot \vec{\dot{x}}\right) = \frac{1}{2}m\left(\dot{x}^2 + \dot{y}^2 + \dot{z}^2\right) \tag{2.19}$$

This provides the three components of a force through kinetic energy Eq. (2.19) as

$$\vec{F} = \frac{d}{dt}\left(\begin{bmatrix} \frac{\partial K}{\partial \dot{x}} & \frac{\partial K}{\partial \dot{y}} & \frac{\partial K}{\partial \dot{z}} \end{bmatrix}^T\right) \tag{2.20}$$

By equating forces we can write Eq. (2.13) for three axes as

$$\frac{d}{dt}\left(\begin{bmatrix} \frac{\partial K}{\partial \dot{x}} & \frac{\partial K}{\partial \dot{y}} & \frac{\partial K}{\partial \dot{z}} \end{bmatrix}^T\right) - \begin{bmatrix} \frac{\partial U}{\partial x} & \frac{\partial U}{\partial y} & \frac{\partial U}{\partial z} \end{bmatrix}^T = 0 \tag{2.21}$$

This gives us Lagrangian formulation of rigid body (particle) for each axis in Cartesian coordinates as

$$\begin{aligned} \frac{d}{dt}\left(\frac{\partial L}{\partial \dot{x}}\right) - \frac{\partial L}{\partial x} = 0 \\ \frac{d}{dt}\left(\frac{\partial L}{\partial \dot{y}}\right) - \frac{\partial L}{\partial y} = 0 \\ \frac{d}{dt}\left(\frac{\partial L}{\partial \dot{z}}\right) - \frac{\partial L}{\partial z} = 0 \end{aligned} \tag{2.22}$$

2.2 Modeling in Cartesian and Spherical/Polar Coordinates

In the above section, we developed a modeling scheme in only a Cartesian coordinate system but often the model is represented in spherical coordinates in 3D space or with polar coordinates in 2D space. The translation and rotational motion also become part of a model in dynamical equations. Now, we discuss the method in 3D space with spherical coordinates, which can also be used in polar coordinates by eliminating an axis. We know that

$$\begin{aligned} x = f(r, \theta, \varphi, t) \\ y = f(r, \theta, \varphi, t) \\ z = f(r, \theta, \varphi, t) \end{aligned} \tag{2.23}$$

We define our system in Cartesian coordinates by using the transformation in Eq. (2.23). Now let us consider that we want to generalize the concept of spherical, polar, or cylindrical coordinates transformed into Cartesian coordinates by using a generalized set of variables instead of r, θ, φ. We now define a generalized vector $\vec{q} = [q_1 \quad q_2 \quad q_3]^T$ and the position in Cartesian coordinates with a vector $\vec{p} = [x \quad y \quad z]^T = [p_1 \quad p_2 \quad p_3]^T$.

$$\begin{bmatrix} p_1 = f(q_1, q_2, q_3, t) \\ p_2 = f(q_1, q_2, q_3, t) \\ p_3 = f(q_1, q_2, q_3, t) \end{bmatrix} = \vec{p} = f\left(\vec{q}, t\right) \tag{2.24}$$

The functions in Eq. (2.24) follow the chain rule for respective derivatives because x, y, z are differentiable with respect to time as well as with respect to each generalized coordinates. Now we follow the convention that $\dot{x} = \frac{dx}{dt}$; a dot on top of the variables represents its differentiation with respect to time for both Cartesian and generalized coordinates. The change with respect to time in Cartesian coordinates is defined as

$$\begin{aligned} \dot{p}_1 &= \frac{\partial p_1}{\partial q_1}\dot{q}_1 + \frac{\partial p_1}{\partial q_2}\dot{q}_2 + \frac{\partial p_1}{\partial q_3}\dot{q}_3 \\ \dot{p}_2 &= \frac{\partial p_2}{\partial q_1}\dot{q}_1 + \frac{\partial p_2}{\partial q_2}\dot{q}_2 + \frac{\partial p_2}{\partial q_3}\dot{q}_3 \\ \dot{p}_3 &= \frac{\partial p_3}{\partial q_1}\dot{q}_1 + \frac{\partial p_3}{\partial q_2}\dot{q}_2 + \frac{\partial p_3}{\partial q_3}\dot{q}_3 \end{aligned} \tag{2.25}$$

Now consider a generalized notation for any element p_i (i.e. x, y, z) of a position vector $\vec{p}$ with respect to time derivates and generalized coordinates

$$\dot{p}_i = \sum_{j=1}^{3} \left(\frac{\partial p_i}{\partial q_j} \dot{q}_j \right) \tag{2.26}$$

The change with respect to generalized coordinates is given as

$$\delta p_i = \sum_{j=1}^{3} \left(\frac{\partial p_i}{\partial q_j} \delta q_j \right) \tag{2.27}$$

We observe that

$$\frac{d}{dt}\left(\dot{p}_i \frac{\partial p_i}{\partial q_j} \right) = \ddot{p}_i \frac{\partial p_i}{\partial q_j} + \dot{p}_i \frac{d}{dt}\left(\frac{\partial p_i}{\partial q_j} \right) \tag{2.28}$$

Solving for the second derivative of position coordinate, we get

$$\ddot{p}_i \frac{\partial p_i}{\partial q_j} = \frac{d}{dt}\left(\dot{p}_i \frac{\partial p_i}{\partial q_j}\right) - \dot{p}_i \frac{d}{dt}\left(\frac{\partial p_i}{\partial q_j}\right) \tag{2.29}$$

Differentiating Eq. (2.26) with respect to q_j where $j = \{1, 2, 3\}$

$$\frac{\partial \dot{p}_i}{\partial \dot{q}_j} = \frac{\partial p_i}{\partial q_j} \tag{2.30}$$

As p_i is a function of q_j then the partial derivatives with respect to time are given as

$$\frac{d}{dt}\left(\frac{\partial p_i}{\partial q_j}\right) = \sum_{k=1}^{3} \frac{\partial}{\partial q_k}\left(\frac{\partial p_i}{\partial q_j}\right) \dot{q}_k \tag{2.31}$$

We know that in complex or real fields, second-order cross partial derivatives are equal for any position variable with respect to other generalized coordinates

$$\frac{\partial}{\partial q_j}\left(\frac{\partial p_i}{\partial q_k}\right) = \frac{\partial}{\partial q_k}\left(\frac{\partial p_i}{\partial q_j}\right) \tag{2.32}$$

Now taking partial derivative of Eq. (2.26) with respect to q_j we get

$$\frac{\partial \dot{p}_i}{\partial q_j} = \sum_{k=1}^{3} \frac{\partial}{\partial q_j}\left(\frac{\partial p_i}{\partial q_k}\right) \dot{q}_k = \sum_{k=1}^{3} \frac{\partial}{\partial q_k}\left(\frac{\partial p_i}{\partial q_j}\right) \dot{q}_k \tag{2.33}$$

By comparing Eqs. (2.31) and (2.33) we get

$$\frac{d}{dt}\left(\frac{\partial p_i}{\partial q_j}\right) = \frac{\partial \dot{p}_i}{\partial q_j} \tag{2.34}$$

Now substituting Eqs. (2.30) and (2.34) in Eq. (2.29)

$$\ddot{p}_i \frac{\partial p_i}{\partial q_j} = \frac{d}{dt}\left(\dot{p}_i \frac{\partial \dot{p}_i}{\partial \dot{q}_j}\right) - \dot{p}_i \frac{\partial \dot{p}_i}{\partial q_j} \tag{2.35}$$

Now we write Eq. (2.35) as

$$\ddot{p}_i \frac{\partial p_i}{\partial q_j} = \frac{d}{dt}\left(\frac{\partial}{\partial \dot{q}_j}\left(\frac{\dot{p}_i^2}{2}\right)\right) - \frac{\partial}{\partial q_j}\left(\left(\frac{\dot{p}_i^2}{2}\right)\right) \tag{2.36}$$

2.3 Work and Energy Formulation

We know that change in work done is the dot product of force applied and corresponding change in position given as

$$\delta W = \vec{F} \cdot \delta \vec{p} \tag{2.37}$$

Equivalently, force vector $\vec{F}$ has three components F_{P_1}, F_{P_2}, and F_{P_3}, such as

$$\delta W = \sum_{i=1}^{3} F_{P_i} \delta_{P_i} \tag{2.38}$$

Each component of force is expressed with Newton's law as

$$F_{P_i} = m\ddot{p}_i \tag{2.39}$$

Equating δW in Eq. (2.38) we have

$$\delta W = \sum_{i=1}^{3} m\ddot{p}_i \delta_{P_i} = \sum_{i=1}^{3} F_{P_i} \delta_{P_i} \tag{2.40}$$

Substituting Eq. (2.27) in Eq. (2.40) we get

$$\delta W = \sum_{j=1}^{3} \left(\sum_{i=1}^{3} m\ddot{p}_i \frac{\partial_{P_i}}{\partial q_j} \right) \delta_{q_j} = \sum_{j=1}^{3} \left(\sum_{i=1}^{3} F_{P_i} \frac{\partial_{P_i}}{\partial_{q_j}} \right) \delta_{q_j} \tag{2.41}$$

Let

$$F_{q_j} = \sum_{i=1}^{3} F_{P_i} \frac{\partial_{P_i}}{\partial_{q_j}} \tag{2.42}$$

Equation (2.41) becomes

$$\delta W = \sum_{j=1}^{3} \left(\sum_{i=1}^{3} m\ddot{p}_i \frac{\partial_{P_i}}{\partial q_j} \right) \delta_{q_j} = \sum_{j=1}^{3} F_{q_j} \delta_{q_j} \tag{2.43}$$

Now by substituting Eq. (2.36) in Eq. (2.43) we get

$$\delta W = \sum_{j=1}^{3} m\left(\sum_{i=1}^{3}\frac{d}{dt}\left(\frac{\partial}{\partial \dot{q}_j}\left(\frac{\dot{p}_i^2}{2}\right)\right) - \frac{\partial}{\partial q_j}\left(\left(\frac{\dot{p}_i^2}{2}\right)\right)\right)\delta_{q_j} = \sum_{j=1}^{3} F_{q_j}\delta_{q_j} \tag{2.44}$$

By definition of kinetic energy of a particle of mass m is given as

$$K = \frac{m}{2}\sum_{i=1}^{3}\dot{p}_i^2 \tag{2.45}$$

There are three Lagrange equations which generate from Eq. (2.44) for each axis ($j = 1, 2, 3$) as

$$m\left(\sum_{i=1}^{3}\frac{d}{dt}\left(\frac{\partial}{\partial \dot{q}_j}\left(\frac{\dot{p}_i^2}{2}\right)\right) - \frac{\partial}{\partial q_j}\left(\left(\frac{\dot{p}_i^2}{2}\right)\right)\right) = F_{q_j} \tag{2.46}$$

Each value of j represents a Lagrange equation in its axis. The purpose of keeping δW in Eq. (2.44) is that we need to equate change in energy through work done by other physical interpretations. The Lagrange equation deals with the kinetic energy of an object in motion in Eq. (2.44). Work done is also represented by change in potential energy as given in Eq. (2.8). There may be non-conservative forces that are acting on the body or internal energy of the system, but due to the law of conservation of energy these forces must balance each other. The negative gradient of potential energy provides forces acting upon the body which may be causing it to move or stop. So if we define potential energy U in Cartesian coordinates, forces in Cartesian coordinates are represented as

$$F_{P_i} = -\frac{\partial U}{\partial p_i} \tag{2.47}$$

Equation (2.42) represent as

$$F_{q_j} = -\sum_{i=1}^{3}\left(\frac{\partial U}{\partial p_i}\right)\frac{\partial_{P_i}}{\partial_{q_j}} = -\frac{\partial U}{\partial q_j} \tag{2.48}$$

Now we write Eq. (2.46) as a Lagrange equation in q_j coordinates

$$m\left(\sum_{i=1}^{3}\frac{d}{dt}\left(\frac{\partial}{\partial \dot{q}_j}\left(\frac{\dot{p}_i^2}{2}\right)\right) - \frac{\partial}{\partial q_j}\left(\left(\frac{\dot{p}_i^2}{2}\right)\right)\right) = -\frac{\partial U}{\partial q_j} \tag{2.49}$$

Again using the definition of kinetic energy from Eq. (2.45) to represent Eq. (2.49) in energy variables

$$\frac{d}{dt}\left(\frac{\partial K}{\partial \dot{q}_j}\right) - \frac{\partial K}{\partial q_j} = -\frac{\partial U}{\partial q_j} \tag{2.50}$$

Kinetic energy is a function of both position and velocity of a particle in generalized coordinate q_j and the potential energy is only a function of generalized position coordinates,

$$\left(\frac{\partial L}{\partial \dot{q}_j}\right) = \left(\frac{\partial K}{\partial \dot{q}_j}\right) \text{ and } \frac{\partial L}{\partial q_j} = \frac{\partial U}{\partial q_j} \tag{2.51}$$

Now using a Lagrange variable L from Eq. (2.14) in three axes system and using Eq. (2.51) in each coordinates we get the following equation for a conservative system in each q_j coordinate

$$\frac{d}{dt}\left(\frac{\partial L}{\partial \dot{q}_j}\right) - \frac{\partial L}{\partial q_j} = 0 \tag{2.52}$$

In a conservative field Eq. (2.7) holds for change in both potential and kinetic energy, but in a non-conservative field the change is also due to non-conservative, external, or internal forces in each coordinate system. E_{nc} is a non-conservative energy due to all of these forces acting upon the body, so the energy equation becomes

$$\Delta K + \Delta U = E_{\text{nc}} \tag{2.53}$$

In this case the Lagrange equation (2.52) does not satisfy the due non-conservative force acting upon a body. Let $F_{\text{nc}q_j}$ is the sum of all non-conservative forces acting in q_j direction. The generalized Lagrange equation in generalized coordinates is

$$\frac{d}{dt}\left(\frac{\partial L}{\partial \dot{q}_j}\right) - \frac{\partial L}{\partial q_j} = F_{\text{nc}q_j} \tag{2.54}$$

Depending upon the space in which coordinates are defined, the value of j can be changed. For a motion with two degrees of freedom there will be only two Lagrange equations but in a motion with three degrees of freedom there will be three Lagrange equations to describe the system. In certain cases, there is motion in both Cartesian coordinates and generalized coordinates; then accordingly a direction of motion in Cartesian is also treated as a variable q_j.

2.4 State Space Representation

The Lagrange equation itself is a nonlinear or linear state space representation. All non-conservative forces are treated as input to the system whether these are controllable or exogenous. The generalized coordinates and their first-order derivatives constitute the state vector.

Example 2.1: A Simple Pendulum The simple pendulum is a body with mass m attached at length l of massless string from the origin (or ground) to move between two end points as shown in Fig. 2.1. If no external force is applied then the movement will depend upon the starting point and eventually comes to end at mass exactly below the origin. In Cartesian coordinates x and y mass move in a semicircle, which can also be related through polar coordinates l and θ, which are fixed length of a string and angle from the origin respectively. There is only one degree of freedom, so there is only one generalized coordinate $q_1 = \theta$. Conventionally these simple examples are represented with θ instead of q_1. Now we define the position and velocity of Cartesian coordinate as a function of a generalized coordinate as

$$\begin{aligned} x(\theta) &= l\ \sin\theta \\ y(\theta) &= -l\ \cos\theta \end{aligned} \tag{2.55}$$

$$\begin{aligned} \dot{x}\left(\theta,\dot{\theta}\right) &= l\dot{\theta}\ \cos\theta \\ \dot{y}\left(\theta,\dot{\theta}\right) &= l\dot{\theta}\ \sin\theta \end{aligned} \tag{2.56}$$

The kinetic energy of the mass m in Cartesian coordinate is given as

$$K = \frac{1}{2}mv^2 = \frac{1}{2}m\left(\dot{x}^2 + \dot{y}^2\right) = \frac{1}{2}ml^2\dot{\theta}^2 \tag{2.57}$$

The potential energy U is due to gravitational pull and it is lowest when $\theta = 0$ or where kinetic energy is highest and vice versa when $\theta = \frac{\pi}{2}$ or $-\frac{\pi}{2}$. So

$$U = mgl(1 - \cos\theta) \tag{2.58}$$

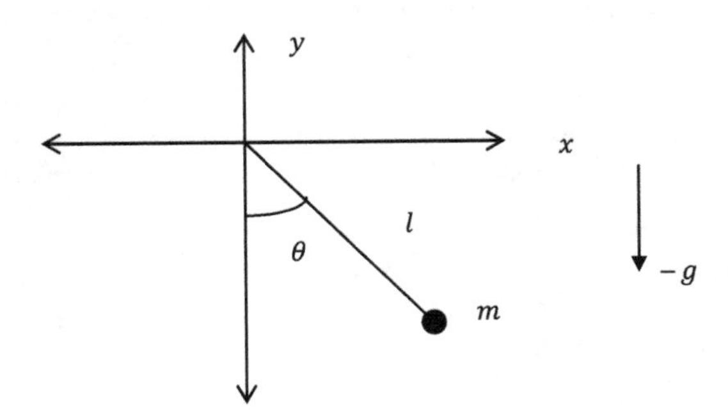

Fig. 2.1 A simple pendulum

The Lagrangian $L = K - U$ is given as

$$L = \frac{1}{2} ml^2 \dot{\theta}^2 - mgl(1 - \cos\theta) \tag{2.59}$$

The Lagrange equation of the system derives as follows

$$\frac{d}{dt}\left(\frac{\partial L}{\partial \dot{\theta}}\right) - \frac{\partial L}{\partial \theta} = 0 \tag{2.60}$$

$$\frac{d}{dt}\left(ml^2\dot{\theta}\right) - mgl\sin\theta = 0 \tag{2.61a}$$

$$\left(ml^2\ddot{\theta}\right) + mgl\sin\theta = 0 \tag{2.61b}$$

The final equation with one degree of freedom now appears as

$$\ddot{\theta} + \frac{g}{l}\sin\theta = 0 \tag{2.62}$$

State space formulation of the system represent Eq. (2.62) with definition of state variables given as $x_1 = \theta$ and $x_2 = \dot{\theta}$. We have state vector $\vec{x} = \begin{bmatrix} \theta \\ \dot{\theta} \end{bmatrix}$ and the nonlinear state space representation as

$$\begin{aligned} \dot{x}_1 &= x_2 \\ \dot{x}_2 &= -\frac{g}{l}\sin x_1 \end{aligned} \tag{2.63}$$

The system can be linearized at relaxed equilibrium point $\begin{bmatrix} \theta_e & \dot{\theta}_e \end{bmatrix}^T = \begin{bmatrix} 0 & 0 \end{bmatrix}$ with only one matrix $A = \begin{bmatrix} 0 & 1 \\ -\frac{g}{l} & 0 \end{bmatrix}$

Example 2.2: Inverted Pendulum on a Moving Cart A cart of mass M is moving by an applied force F acting along x-axis. An inverted pendulum of mass m is attached on the center of cart with a rod of length l (Fig. 2.2).

The motion of the cart is defined from origin and the angle is measured from vertical y-axis. The system can either be represented with horizontal and vertical variables of cart and bob or by using horizontal displacement of cart and angular position of bob. There are only two degrees of freedom; an angle can be measured from the horizontal but in order to be conversant with remaining literature we chose the angle θ measured from y-axis. So for 2-DoF system, there are two generalized coordinates $x_1 = x$, the cart's displacement and θ the angle of a bob from vertical. The position of bob (x_2, y_2) can be given in terms of x_1 and θ as

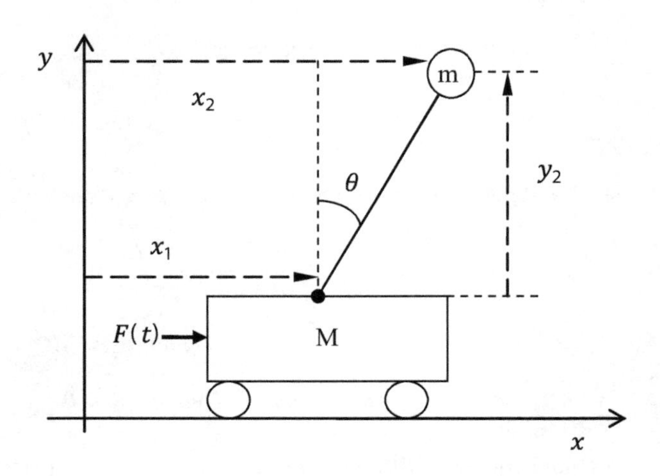

Fig. 2.2 Inverted pendulum on a moving cart

$$\begin{aligned} x_2 &= x + l \cdot \sin\theta \\ y_2 &= l \cdot \cos\theta \end{aligned} \tag{2.64}$$

The velocity components of bob are given as

$$\begin{aligned} \dot{x}_2 &= \dot{x} + l \cdot \dot{\theta} \cdot \cos\theta \\ \dot{y}_2 &= -l \cdot \dot{\theta} \cdot \sin\theta \end{aligned} \tag{2.65}$$

The kinetic energies of both masses are given as

$$K_1 = \frac{1}{2} M \dot{x}^2 \tag{2.66}$$

$$K_2 = \frac{1}{2} m\left(\dot{x}_2{}^2 + \dot{y}_2{}^2\right) \tag{2.67a}$$

$$K_2 = \frac{1}{2} m\left[\left(\dot{x} + l \cdot \dot{\theta} \cdot \cos\theta\right)^2 + \left(-l \cdot \dot{\theta} \cdot \sin\theta\right)^2\right] \tag{2.67b}$$

$$K_2 = \frac{1}{2} m\left(\dot{x}^2 + 2l\dot{x}\dot{\theta} \cdot \cos\theta + l^2\dot{\theta}^2\right) \tag{2.67c}$$

So the total kinetic energy of the system from Eqs. (2.66) and (2.67c) is given as

$$K = K_1 + K_2 = \frac{1}{2}(M + m)\dot{x}^2 + ml\dot{x}\dot{\theta} \cdot \cos\theta + \frac{1}{2} ml^2\dot{\theta}^2 \tag{2.68}$$

The potential energy in a bob is given as

$$U = mgy_2 = mgl\cos\theta \tag{2.69}$$

The Lagrangian L is given as

$$L = K - U = \frac{1}{2}(M+m)\dot{x}^2 + ml\dot{x}\dot{\theta} \cdot \cos\theta + \frac{1}{2}ml^2\dot{\theta}^2 - mgl\cos\theta \tag{2.70}$$

we represent two equations of motion sing the variable L and two generalized coordinates (x, θ) as

$$\frac{d}{dt}\left(\frac{\partial L}{\partial \dot{x}}\right) - \frac{\partial L}{\partial x} = F \tag{2.71}$$

and

$$\frac{d}{dt}\left(\frac{\partial L}{\partial \dot{\theta}}\right) - \frac{\partial L}{\partial \theta} = 0 \tag{2.72}$$

From Eqs. (2.70), (2.71), and (2.72) are solved as

$$\frac{\partial L}{\partial \dot{x}} = (M+m)\dot{x} + ml\dot{\theta} \cdot \cos\theta \tag{2.73a}$$

$$\frac{\partial L}{\partial \dot{\theta}} = ml\dot{x} \cdot \cos\theta + ml^2\dot{\theta} \tag{2.73b}$$

$$\frac{\partial L}{\partial x} = 0 \tag{2.73c}$$

$$\frac{\partial L}{\partial \theta} = mgl\sin\theta \tag{2.74}$$

So equations of motion are

$$(M+m)\ddot{x} + ml\ddot{\theta} \cdot \cos\theta - ml\dot{\theta}^2\sin\theta = F \tag{2.75}$$

$$ml\ddot{x} \cdot \cos\theta - ml\dot{x}\dot{\theta}\sin\theta + ml^2\ddot{\theta} - mgl\sin\theta = 0 \tag{2.76}$$

These are nonlinear equations for the system of a moving cart with an inverted pendulum. In order to make a state space system, first we need to solve variables with double derivatives simultaneously. Each generalized coordinate of the system is second-order so a total order of the system is 4. We have both $\ddot{x}$ and $\ddot{\theta}$, so we need to solve these equations simultaneously. Eqs. (2.75) and (2.76) are represented as

$$ml^2\ddot{\theta} = \left(\frac{l}{\cos\theta}\right)\left[F - (M+m)\ddot{x} + ml\dot{\theta}^2\sin\theta\right] \tag{2.77}$$

$$\ddot{x} = \dot{x}\dot{\theta}\tan\theta - \frac{l\ddot{\theta}}{\cos\theta} + g\tan\theta \tag{2.78}$$

Inserting value of $\ddot{x}$ from Eq. (2.78) in Eq. (2.75) and then inserting the value of $ml^2\ddot{\theta}$ from Eq. (2.77) in Eq. (2.76) simultaneously solve the equations for state space representation of a state vector $\begin{bmatrix} x & \theta & \dot{x} & \dot{\theta} \end{bmatrix}^T$ as

$$\ddot{x} = \left[m\cos\theta - \left(\frac{M+m}{\cos\theta}\right)\right]^{-1}\left(m\dot{x}\dot{\theta}\sin\theta + mg\sin\theta + ml\dot{\theta}^2\tan\theta + \frac{F}{\cos\theta}\right) \tag{2.79}$$

$$\ddot{\theta} = \left(\frac{1}{l}\right) \times \left[-\frac{M}{\cos\theta} - \frac{m}{\cos\theta} + m\cos\theta\right]^{-1}\left(-(M+m)\left(\dot{x}\dot{\theta}\tan\theta + g\tan\theta\right) + ml\dot{\theta}^2\sin\theta + F\right) \tag{2.80}$$

Problems

P2.1 Consider spherical coordinates of mass m in motion

$$x = r\cos\theta\sin\varphi$$
$$y = r\sin\theta\sin\varphi$$
$$z = r\cos\varphi$$

Formulate Lagrangian equation of motion in three axis for three generalized coordinates and represent as state space model.

P2.2 A pendulum given in Example 2.1 is rotating with angular velocity ω in a circular path of radius r. The motion is under the influence of gravity with no external forces acting upon it. The Cartesian coordinates are given as

$$x(t) = r\cos(\omega t) + l\cos\theta$$
$$y(t) = -r\sin(\omega t) + l\sin\theta$$

Find the state space representation of a system through Lagrangian formulation.

P2.3 Define a generalized coordinate system for the problem given in Example 2.2 and represent its nonlinear state space formulation. Use the symbolic math toolbox to obtain Jacobian matrices A and B given in Eq. (1.19) and Eq. (1.20) by linearizing at relaxed position.

P2.4 Nonlinear systems are also linearized by approximating nonlinear terms within a specified range on equilibrium point. The inverted pendulum on a cart can be linearized by approximating a small angle and negligible velocities at equilibrium point for trigonometric functions i.e., $\sin\theta \approx \theta$, $\cos\theta \approx 1$. Find a linear state space equation by these assumptions and obtain a state space representation of the inverted pendulum on a cart system.

P2.5 The position of bob of the inverted pendulum on a moving cart can be controlled by applying external torque τ at the hinge changing the Eq. (2.78) as

$$ml\ddot{x}\cdot\cos\theta - ml\dot{x}\dot{\theta}\sin\theta + ml^2\ddot{\theta} - mgl\sin\theta = \tau$$

Linearize the complete system by the assumption given in P2.4 and obtain state space representation to monitor all state variables at output independently.

References

1. Friedland, Bernard. 2005. *Control System Design—An Introduction to State Space Methods*. Mineola: Dover Publications.
2. Zak, Stanislaw H. 2002. *Systems and Control*. Oxford: Oxford University Press.
3. Brizard, Alain J. 2008. *An Introduction to Lagrangian Mechanics*. Singapore: World Scientific Publishing.
4. Wellstead, P.E. 1979. *Introduction to Physical System Modelling*. London: Academic Press.
5. Amirouche, Farid. 2006. *Fundementals of Multibody Dynamics—Theory and Applications*. Basel: Birkhäuser.

Chapter 3
Introduction to Bond Graph Modeling

Dynamical systems are classified by the form of energy consumed or delivered, e.g., electrical, mechanical, chemical, and hydraulic systems. There are different physical laws dealing with different energy systems and therefore different modeling schemes formulate the differential equations accordingly. Universally there is only one source of energy behind all sources as our understanding, but modern day's physics is still not able to prove the unification of all forces together. The law of conversation of energy has the same units of energy consumption as well as the rate of change of energy in all forms. This law helps to form a unified modeling approach for different types of energy systems. Modeling of an electromechanical system requires modeling through electrical laws as well as mechanical laws, with governing equations between energy transfers. Some examples are electromagnetic relays, servo control hydraulic systems, pressure valves, and many other similar systems. In the 1960s a professor named Henry M. Paynter of MIT introduced this scheme by modeling through power transformation. A graphical bond represents the power flow or transformation from one end to other end. A bond represents a single power transformation between two ports or ends. The interconnection of these bonds leads us to develop a complete model of a required system. A half arrow is a graphical representation to show the direction of power flow in the model as shown in Fig. 3.1.

3.1 Power Variables

The bond graph methodology works on analogy of power variables in different energy domains. In every domain there are two variables whose product is power measured in Watts or rate of change of energy J/s. One of these variables is called effort $e(t)$ and the other variable is called flow $f(t)$. In most cases effort is a cause that is exciting the system and flow is an effect or response in the system. In bond graph terminology we called effort and flow variables, but we recognize these

A.M. Mughal, *Real Time Modeling, Simulation and Control of Dynamical Systems*,
DOI 10.1007/978-3-319-33906-1_3

Fig. 3.1 A half arrow representing a bond between two energy ports

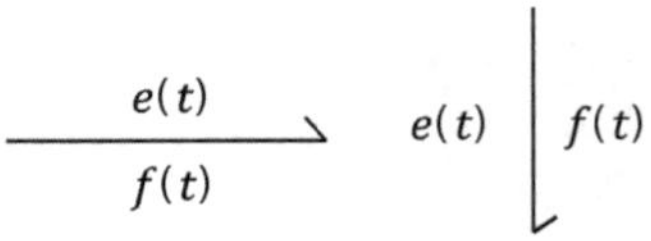

$e(t)$
$f(t)$
$e(t)$ $f(t)$

Fig. 3.2 Bond graph representation with power

Table 3.1 Effort and flow variables in different energy systems

Energy system		Effort	Flow
Electrical		Voltage (V)	Current (A)
Magnetic		Magnetomotive force (V)	Flux rate (W/s)
Mechanical	Translational	Force (N)	Velocity (m/s)
	Rotational	Torque (Nm)	Angular velocity (rad/s)
Hydraulics		Pressure (N/m^2)	Volume flow rate (m^3/s)
Thermal		Pressure (N/m^2)	Volume change rate (m^3/s)
		Temperature (K)	Entropy Change Rate (J/K/s)
		Temperature (K)	Heat Change rate
Chemical		Enthalpy	Mass flow rate (kg/s)
		Chemical Potential (J/mol)	Mole flow rate (mol/s)

variables in each energy domain with different names. The product of effort and flow is power $P(t)$ in every energy domain.

$$P(t) = e(t) \cdot f(t) \tag{3.1}$$

In bond graph representation shown in Fig. 3.2, the effort variable is written above or to the left and the flow variable is written in under or to the right side of the half arrow.

In electrical systems, effort is voltage and flow is current and their product is power either in watts, VA or VAR, each of which is a representation of power in Watts or Joules per second. In mechanical energy for translational motion, effort is force and flow is velocity and their product is power either in Watts or horsepower (hp). Conceptually voltage is the driving force in electrical systems and force is the excitation in mechanical systems. Current flows in the electrical circuits and a mechanical body moves with a velocity. Similarly, we understand efforts and flow variables for other types of energy systems. Table 3.1 shows the effort and flow variables for different energy systems

3.2 Pseudo Bond Graph

In some cases it is not possible to generate a set of power variables whose product is power. In Table 3.1, one set of power variables for a thermal system shows temperature and heat change rate as effort and flow variables respectively. The product is not power rather power times temperature, which conveys that a power transfer is taking place through change in temperature. This type of bond where product of effort and flow is not purely power is called pseudo bond graph.

Different effort and flow variables can be associated with the same energy systems depending upon their application and integration with other systems. Each type of energy systems have already established modeling techniques through their respective laws of physics, but a bond graph modeling scheme is vital when more than one energy systems are modeled together.

3.3 Energy Variables

Power is the rate of change of energy in the system, so the change must be occurring from either effort or flow and in certain cases from both variables. There are two variables that associate effort and flow variables with change in the energy of the system. These variables are called energy variables. One variable is called generalized momentum $p(t)$ and change in momentum is an effort at the port. Another variable is generalized displacement $q(t)$ and the change in generalized displacement is the flow variable. In electrical systems, change in flux linkage is voltage/effort so flux linkage $\lambda(t)$ is the generalized momentum in electrical systems. The rate of change of electrical charge is the current flowing in the conductor, which provides us generalized displacement as electrical charge $q(t)$. In a mechanical system, momentum of motion (translational or rotational) is the generalized momentum. As we know, mechanical momentum for translational motion is the product of mass and velocity given as $p(t) = m \cdot v(t)$ and change in momentum $\dot{p}(t) = m \cdot \dot{v}(t) = m \cdot a(t) = F(t) = e(t)$ is a respective force. Similarly, change in angular momentum is the torque, which is an effort variable in rigid body rotational motion. The generalized effort and generalized momentum relationship is defined as

$$\dot{p}(t) = e(t) \tag{3.2a}$$

$$p(t) = \int_0^t e(t)dt \tag{3.2b}$$

Similarly the relationship of generalized flow and generalized displacement is defined as

$$\dot{q}(t) = f(t) \tag{3.3a}$$

$$q(t) = \int_0^t f(t)dt \tag{3.3b}$$

As we know that power is rate of change of energy,

$$P(t) = \frac{dE(t)}{dt} \tag{3.4a}$$

$$E(t) = \int_0^t P(t)dt \tag{3.4b}$$

Substituting Eq. (3.1) yields the definition of energy function as

$$E(t) = \int_0^t e(t) \cdot f(t)\, dt \tag{3.5}$$

Now we can define energy function with either of the energy variables by substituting the relationship of effort Eqs. (3.2a and 3.2b) or flow Eqs. (3.3a and 3.3b) in Eq. (3.5). By substituting Eq. (3.2a) in Eq. (3.5), we get energy function as a flow variable integrated with momentum.

$$E(t) = \int_0^t f(t) \cdot dp(t) \tag{3.6a}$$

$$E(t) = \int_0^p f(p) \cdot dp \tag{3.6b}$$

Using the relationship of generalized flow and displacement energy from Eqs. (3.3a and 3.3b) for Eq. (3.5) give us energy function as an effort integrated over displacement.

$$E(t) = \int_0^t e(t) \cdot dq(t) \tag{3.7a}$$

$$E(t) = \int_0^q e(q) \cdot dq \tag{3.7b}$$

Both Eqs. (3.6a and 3.6b) and (3.7a and 3.7b) represent energy $E(t)$ as a function of time, whereas momentum and displacement are autonomous functions of time. Power is a product of flow and effort variables whereas energy is an integral of the power. The energy is obtained by integrating either effort variable with respect to displacement or flow variable with respect to momentum. So in a bond there are actually four variables related together; two are power variables, which are integrals of their respective energy variables. In many cases both effort and flow are often not measureable, and power is determined through integration of momentum or displacement variables. This relationship was explained as a tetrahedron of states

Table 3.2 Energy variables for different systems

Energy system		Momentum $p(t)$	Displacement $q(t)$
Electrical		Flux linkage (V s)	Charge (C = A s)
Magnetic		Flux linkage (V s)	Flux (Wb)
Mechanical	Translational	Momentum (N s)	Displacement (m)
	Rotational	Angular momentum (N m s)	Angle (rad)
Hydraulics		Pressure momentum (N s/m^2)	Volume (m^3)
Thermal		Pressure (Pa)	Volume (m^3)
		Temperature (K)	Entropy (J/K)
		Temperature (K)	Heat (J)
Chemical		Enthalpy (J/kg)	Mass flow (kg)
			Mole (moles)

between power and energy variables in [1]. Energy function in Eqs. (3.6a and 3.6b) and Eqs. (3.7a and 3.7b) is an accumulation of integration at any instantaneous value, so both energy variables are associated with energy-storing elements of a physical system. Energy variable is an integral of a power variable and stores energy with other power variables for different systems shown in Table 3.2.

3.4 Excitation and Response

The input to the system is excitation or cause to excite the model which produces effect or response as output. The half arrow representation of a bond determines a power input, which is flowing through one end to half arrow end. The source of power attached with one end provides power to the other end of system (half arrow side) conventionally given in Table 3.3. The relationship from output (effect, response) to input (cause, excitation) is called transfer function. In real-time modeling, input is excited with impulse, step or at different frequencies to check the dynamic response of the output.

3.5 Word Bond Graph

Word bond graph is a first step to represent the block diagrams of a model towards complete bond graph modeling. These bond graphs explain the structure of the system, and the flow of power and interconnectivity of the components. These are like signal flow diagrams of the electrical circuits or free body diagrams. We first draw an electromechanical system as word bond graph and later expand it into a full structured bond graph model. The word bond graphs of various energy ports are given in Table 3.4

Table 3.3 Sources/excitation of different systems with word bond graphs

Sources/excitation	Figure	Word bond graph
Electric voltage sources		$V(t)$
Electric current source		$i(t)$
Force	F	$F(t)$
Torque	T	$\tau(t)$
Relative velocity actuator	v	$v(t)$

3.6 Causality Assignment

It is important to understand that the power variable is the product of effort and flow variables, which means either effort or flow is going into the bond (port) or coming out of the port. When two components are connected through a bond, if one component is taking effort as input then it must be giving flow as output or vice versa. Causal stroke determines the output of bond as either effort or flow variable. Conventionally we place causal stroke at the side of the component, which is providing flow to the other component. So causal stroke end specifies where flow is being determined. Assigning causality through a stroke helps to write algebraic relationships between ports by determining which port is providing effort or flow. We know that the law of conservation of energy forms the basic principle of bond graph modeling by equating powers at both sides. Accordingly, if one variable is known at one end, then the other end measures other power variables.

Two systems, A and B, interconnect with equal power at both ends as shown in Fig. 3.3, so there are a variety of ways in determining effort, flow, and power transfer between two systems. There are two choices for power transfer and two choices for assigning causality through a stroke at each end. Overall there are four possible representations with respective understanding as given in Table 3.5.

3.7 Active Bond

If we connect two elements and power transfer is not taking place, but either effort or flow is providing a transfer, then this action is represented by active bond. Active bond shown in Fig. 3.4 is denoted by a full arrow with either effort or flow marked as a main variable and the other variable assumed as zero. In reality, there is a very low power being transmitted but cannot be equated with the other end and preceding systems. This situation arises when some other power system is driving some

Table 3.4 Word bond graph of different energy ports

Name	Figure	Word bond graph
Electric motor		$V(t)$ / $i(t)$ → Motor → $\tau(t)$ / $\omega(t)$
Belt pulley		$\tau(t)$ / $\omega(t)$ → Belt Pulley → $F(t)$ / $v(t)$
Cam rod		$\tau(t)$ / $\omega(t)$ → Cam Rod → $F(t)$ / $v(t)$
Crank rod		$F(t)$ / $v(t)$ → Crank Rod → $\tau(t)$ / $\omega(t)$
Rack and pinion (fixed pinion)		$F(t)$ / $v(t)$ → Rack Pinion → $\tau(t)$ / $\omega(t)$
Rack and pinion (fixed rack)		$\tau(t)$ / $\omega(t)$ → Rack Pinion → $F(t)$ / $v(t)$
Spindle		$\tau(t)$ / $\omega(t)$ → Spindle → $F(t)$ / $v(t)$
Lever		$F_1(t)$ / $v_1(t)$ → Lever → $F_2(t)$ / $v_2(t)$
Electrical transformer	n : 1	$V_1(t)$ / $i_1(t)$ → Transformer → $V_2(t)$ / $i_2(t)$
Universal coupling		$\tau_1(t)$ / $\omega_1(t)$ → Cam Rod → $\tau_2(t)$ / $\omega_2(t)$

Fig. 3.3 Interconnection of two systems

components and thus represented by active bond. A voltage amplifier takes voltage input and then converts it by multiplying with a fixed gain value. The power at the input of the amplifier is not equal to the power at the output because the

Table 3.5 Causality and power transfer between two systems

Representation	Explanation
A —$e(t)$ / $f(t)$→| B	A is supplying power to B, Output of A is effort, Input to A is flow, Output of B is flow, and Input to B is effort $f(t) = \Phi\{e(t)\}$
A |—$e(t)$ / $f(t)$→ B	A is supplying power to B, Output of A is flow, Input to A is effort, Output of B is effort, and Input to B is flow $e(t) = \Phi\{f(t)\}$
A ←$e(t)$ / $f(t)$—| B	B is supplying power to A, Output of A is effort, Input to A is flow, Output of B is flow, and Input to B is effort $e(t) = \Phi\{f(t)\}$
A |←$e(t)$ / $f(t)$— B	B is supplying power to A, Output of A is flow, Input to A is effort, Output of B is effort, and Input to B is flow $f(t) = \Phi\{e(t)\}$

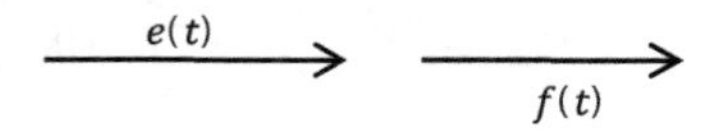

Fig. 3.4 Active Bond Graphs, (**a**) Active Effort and flow is zero, (**b**) Active flow and effort is zero

Table 3.6 Active bonds of different models

Name	Figure	Word bond graph
Operational amplifier	− +	$V_1(t)$ / $i_1(t)$ → Op Amp → $V_2(t)$
Controlled voltage source	+	$V_1(t)$ / $i_1(t)$ → CVS → $V_2(t)$
Controlled current source		$V_1(t)$ / $i_1(t)$ → CCS → $i_2(t)$
Relative force actuator	← F →	↓ $F(t)$ →
Sensor (any power or energy variable)	P	$e(t)$ / $f(t)$ → Sensor → Req. Variable

amplification is also receiving power from another source as shown in Table 3.6 for different sources. Graphically, we represent active bond without any causality assignment; either effort or flow is the key input through this bond and the other variable is assumed at very low value and approximated as zero.

3.8 Connection of Bonds

The bonds or ports are connected with each other and thus form a complete bond graph. A block diagram representation followed by a word bond graph is usually the first step. At each bond, power is equal to the proceeding or succeeding bond, and thus complete variables can be solved.

Example 3.1: Bond Graph of Multienergy System Construct a word bond graph of the block diagram shown in the system of an overhead storage tank filled through electric power by a driving shaft coupled with AC motor and hydraulics pump. Calculate the following

(a) What is the driving current if the system is consuming 1 kW of power?
(b) What is the constant pressure at the pump when a 400 L tank fills in an hour?
(c) Drive shaft is reducing the speed of motor by one-tenth, and torque coupling of the shaft with AC motor is 5 Nm/A. What is torque and the speed of drive shaft at the pump?
(d) Drive a numeric relation between pressure of the pump and the torque of the drive shaft (Fig. 3.5).

Solution A simply connected word bond graph of a system is given in Fig. 3.6, it shows the coupling of the interconnected systems:

(a) In a AC 220 V system, driving current of the motor is

$$i(t) = \frac{P(t)}{V(t)} \tag{3.8a}$$

$$i(t) = \frac{1000}{200} = 4.55 \text{ A} \tag{3.8b}$$

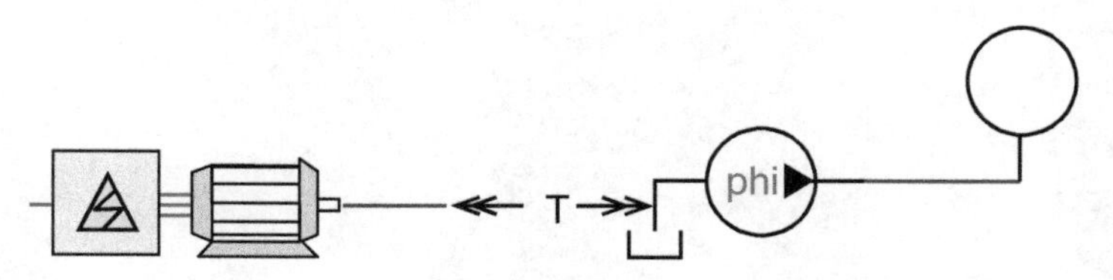

Fig. 3.5 Block diagram of a water storage system. AC motor driving a shaft, a shaft is rotating a pump which is storing water in over head tank

Power Source $\xrightarrow[i(t)]{V(t)}$ Electric Motor $\xrightarrow[\omega_1(t)]{\tau_1(t)}$ Drive Shaft $\xrightarrow[\omega_2(t)]{\tau_2(t)}$ Pump $\xrightarrow[Q(t)]{P(t)}$ Volume Sensor

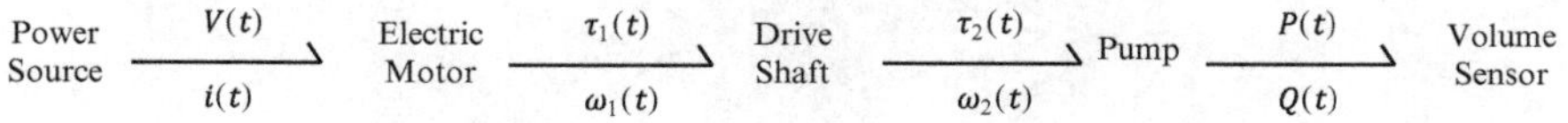

Fig. 3.6 Word bond graph of a water storage system. AC motor driving a shaft, a shaft is rotating a pump which is storing water in over head tank

(b) Volume flow rate is calculated as

$$\dot{Q} = \frac{\Delta \text{Vol}}{\Delta \text{time}} = \frac{0.4 \text{ m}^3}{3600 \text{ sec}} = 11.1 \times 10^{-3} \, {}^{\text{m}^3}/_{\text{sec}} \tag{3.9}$$

The pressure is a generalized effort calculated as

$$P_{\text{hyd}} = \frac{P}{\dot{Q}} = \frac{1000}{11.1 \times 10^{-3}} = 9 \text{ M Pa} \tag{3.10}$$

(c) The torque of drive shaft is linked with current of the motor as

$$\tau_1(t) = 5 \cdot i(t) = 22.73 \text{ Nm} \tag{3.11}$$

This implies the speed of motor is

$$\omega_1(t) = \frac{P}{\tau_1} = \frac{1000}{22.73} = 44 \text{ rad/ sec} \tag{3.12}$$

The drive shaft at the pump is rotating as

$$\omega_2(t) = \frac{\omega_1(t)}{10} = 4.4^{\text{rad/ sec}} \tag{3.13}$$

The torque at the pump end is now calculated as

$$\tau_2(t) = \frac{P}{\omega_2} = \frac{1000}{4.4} = 227.27 \text{ Nm} \tag{3.14}$$

(d) The pressure of the pump and the torque of the motor is related as power variables

$$P_{\text{hyd}} = \tau_1 \cdot \frac{\omega_1}{\dot{Q}} \tag{3.15}$$

Problems

P3.1 Draw the word bond graph of the following systems

Name	Figure
Gear	
Timing Belt	
Thermistor	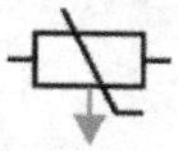
Tachometer/encoder	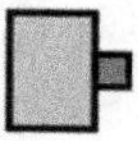
Fork (a combination of sources)	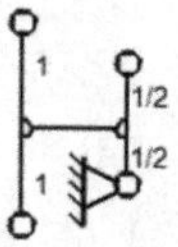
Heat flow sensor	

P3.2 A 50 V dc motor is attached to a pulley of 1 m diameter, which is driving a bucket of mass 20 kg vertically upwards 50 m by consuming a current of 2 A in an earth gravity field ($g = 10$ m/s^2).

(a) How long does the motor have to work to pull the bucket 50 m upwards?
(b) What will be the linear velocity of the bucket in vertical direction?
(c) Draw a word bond graph for the system.
(d) What torque will the motor shaft experience?
(e) How many revolutions of the pulley are needed?
(f) Is there any associated energy variable at any port which can easily be determined?
(g) Represent the complete system as a single two-port element with appropriate ratio.

References

1. Karnopp, Dean C., Donald L. Margolis, and Ronald C. Rosenberg. 2012. *System Dynamics—Modeling and Simulation of Mechatronics Systems*, 5th ed. Hoboken: Wiley.
2. Borutzky, Wolfgang. 2010. *Bond Graph Methodology—Development and Analysis of Multidisciplinary Dynamic System Modeling*. London: Springer.

Chapter 4
Elements of Bond Graph

Components and elements bond the system, which define the relationship between two consecutive ports. These bonding elements follow constitutive laws for equating power between bonds. These elements are analogues in different energy systems as we discussed effort and flow variables in Chap. 3. If a bond or port has only one element at either end and does not require any other element to define it, then it is a 1-port element. If an element requires two bonds for defining relationship, then it is a 2-port element. If three or more bonds connect for a constitutive law, then these elements are called junctions. It is important to note that these 1-port and 2-port elements and junctions are different than vector bond graphs, which will be discussed later.

4.1 1-Port Elements

There are five 1-port elements, two of which are sources, one of which is dissipative, and two of which are energy storing elements.

4.1.1 Sources

Source of generalized effort and source of generalized flow are both 1-port elements. Voltage sources in electrical systems and applied force or applied torque in mechanical systems are sources of effort. Graphically, source of effort connects at the starting end of the bond with a causal stroke on the other side (Fig. 4.1).

Generally, sources provide power to the systems, so the half arrow side is away from the elements where source is placed. Similarly, source of flow represents the current source in an electrical system or the source of fixed velocity in a mechanical

A.M. Mughal, *Real Time Modeling, Simulation and Control of Dynamical Systems*,
DOI 10.1007/978-3-319-33906-1_4

Se ——⇁|

Fig. 4.1 Bond graph of a generalized source of effort

Sf |——⇁

Fig. 4.2 Bond graph of a generalized source of flow

system. Figure 4.2 represents the source of generalized flow with causality assignment near the source.

In source of effort, effort is fixed and flow is arbitrary in the bond; in source of flow, flow is fixed and effort is arbitrary in the bond. Causality assignments for both sources are fixed and cannot be changed as flow is uniquely determined away from S_e and at the source in S_f. In different literature, it is often confusing when sources are mentioned particularly for energy systems, for example S_F is also mentioned as a source of force, which is an effort variable but also as a source of flow. S_V is often mentioned for the source of velocity, which is flow variable as well as for source of voltage, which is an effort in electrical systems. In this text, S_e is a generalized source of effort that represents source of voltage, force, torque, etc in all energy systems. Similarly, source flow S_f represents current, velocity, angular velocity or volume flow rate, etc. A source in bond graph represents the ideal source with a given value whether constant or alternating, sinusoidal, decaying or time-based function. If nonlinearities, dissipation, or dependencies upon other power variables is to be considered then these sources can be modeled as a complete system. For example, a 12-V voltage source of a car battery with power rating of 100 Ah is dependent upon the current supplied. In a bond graph model, a simple S_E with assigned value of 12 V will model the system but it will not take the power consideration of 100 Ah in the simulation. In order to represent a non-ideal source, a detail bond graph model for source is required.

4.1.2 1-Port Resistor

Resistor is an element that defines the relationship between generalized effort and generalized flow. Nonlinear constitutive law of a resistor is given as

$$e(t) = \varphi_R\{f(t)\} \tag{4.1}$$

or

$$f(t) = \varphi_R^{-1}\{e(t)\} \tag{4.2}$$

Resistor represents a direct relationship between both power variables. In electrical systems, it is a simple resistance in ohms (Ω). Resistance is energy dissipating

Table 4.1 Bond graph of 1-port resistor with constitutive law

Bond graph	Constitutive law
⊢⇀ R	$e(t) = R \cdot f(t)$
⇀⊣ R	$f(t) = \frac{e(t)}{R}$

element which is defined by a direct proportion between effort and flow. Linearly these relations are given as

$$e(t) = R \cdot f(t) \tag{4.3}$$

or

$$f(t) = \frac{e(t)}{R} \tag{4.4}$$

In mechanical systems, damping $b\,(\mathrm{N\ s\ m^{-1}})$ is defined with a relationship between force and velocity given as $F(t) = b \cdot v(t)$. In mechanical rotation it is a rotational damping c (N m s) between torque and angular velocity given as $\tau(t) = c \cdot \omega(t)$. Generalized resistance is represented as R with appropriate units of the given system. Graphically, bond graph of generalized resistor is given in Table 4.1 with two possibilities of causality assignment.

Constitutive relationships at the resistor end determine the effort or flow variable. Either effort is determined if causal stroke is away from it or flow is determined if causal stroke is towards its end. Resistance is an energy dissipating element and appears at the half arrow side, and there is no preference for defining causal stroke on either ends.

4.1.3 1-Port Capacitor

The remaining two 1-port elements are energy storing elements, which are important in order to obtain differential equations for the system. These two elements define a relationship between a power variable to an energy variable. The capacitor is a first 1-port energy storing element which directly holds a relationship between generalized effort and generalized displacement. As generalized displacement is integral of generalized flow then the relationship between effort and flow of the bond with capacitor has an integral or derivative given as

$$e(t) = \varphi_{\mathrm{C}}^{-1}\left\{\int_0^t f(t) \cdot dt\right\} \tag{4.5}$$

$$f(t) = \frac{d}{dt}\varphi_{\mathrm{C}}\{e(t)\} \tag{4.6}$$

Table 4.2 Bond graph of 1-port capacitor with constitutive law

⊢⇀ C	$e(t) = \frac{1}{C}\left\{\int_0^t f(t)\cdot dt\right\}$
⇀| C	$f(t) = C\cdot\frac{de(t)}{dt}$

Linearly these relationships are given as

$$e(t) = \frac{q(t)}{C} = \frac{1}{C}\left\{\int_0^t f(t)\cdot dt\right\} \tag{4.7}$$

or

$$q(t) = C\cdot e(t) \tag{4.8}$$

This implies that

$$f(t) = C\cdot \dot{e}\,(t) \tag{4.9}$$

It is clear from expression that if effort is given then flow is arbitrarily determined or vice versa and thus affects the causality assignment accordingly as given in Table 4.2. Generalized capacitor is preferred with integral causality capacitor over differential causality.

In electrical systems, the capacitor is capacitance {in Farads} defined through the relationship of applied voltage and the charge developed by relation $q(t) = C\cdot V(t)$. In mechanical translational systems, the capacitor is spring constant k $\{\mathrm{N\ m^{-1}}\}$ by applied force and displacement by which spring compresses or stretches by defining relation $F(t) = k\cdot x(t)$. By the concept of generalized capacitance, spring constant k is known as a stiffness parameter, which is the inverse relation as compared to electrical systems. However, in rotational mechanics, both rotational spring constant k $\{\mathrm{N\ m\ rad^{-1}}\}$ and torsional constant c $\{\mathrm{rad\ N^{-1}\ m^{-1}}\}$ are used.

4.1.4 1-Port Inertia

The relationship between generalized momentum and generalized flow defines inertia at the bond. It is also an energy storing element, unlike the capacitor which stores generalized efforts, it stores generalized flow. Integral of effort is momentum of the system and the relationship between power variables with inertia at the one end of the bond is given as

$$f(t) = \varphi_{\mathrm{I}}^{-1}\left\{\int_0^t e(t)\cdot dt\right\} \tag{4.10}$$

Table 4.3 Bond graph of 1-port inertia with constitutive law

Bond graph	Constitutive law
⊢——⇀ I	$e(t) = I \cdot \frac{df(t)}{dt}$
——⇀⊣ I	$f(t) = \frac{1}{I} \cdot \left\{ \int_0^t e(t) \cdot dt \right\}$

$$e(t) = \frac{d}{dt} \varphi_{\mathrm{I}}\{f(t)\} \tag{4.11}$$

Linearly these relationships are given as

$$f(t) = \frac{p(t)}{I} = \frac{1}{I} \left\{ \int_0^t e(t) \cdot dt \right\} \tag{4.12}$$

or

$$p(t) = I \cdot f(t) \tag{4.13}$$

Equivalently

$$e(t) = I \cdot \dot{f}(t) \tag{4.14}$$

Causality assignment for generalized inertia is demonstrated in Table 4.3

In electrical systems, inertia is an inductance L {Henry} by a defining relationship between voltage applied and the current passing through the inductor through flux linkage (which is momentum in electrical systems) given as $\lambda(t) = L \cdot i(t)$. In a mechanical system, the term inertia is more familiar, as in mechanical translational systems, generalized inertia is referred as mass of the body which has defining relation between momentum and velocity as $p(t) = m \cdot v(t)$. Similarly in a mechanical rotational system, angular momentum and angular velocity is defined by an inertia of the system as $p_\tau = J \cdot \omega(t)$.

4.2 Derivative (Differential) Causality and Integral Causality

In case of energy storing ports, the concept of derivative (differential) and integral causalities prompts an important step in modeling. Algebraically, effort or flow can be determined at either side of bond but in modeling it requires a preference to integral causality over derivative causality. In the case of sources, causality assignment is fixed as either effort or flow determines at respective source end and hence its causality cannot change. In the case of resistor, which is an energy dissipating element, both causality assignments make an algebraic relationship between both power variables. Both choices are fine in modeling because it does not induce or deduce existing information of the model. In the case of capacitor and inertia, defining relationship between effort and flow is either an integral or derivative of one variable to determine other variable. In the case of integral causality, it means existing/past information is being added to formulate a value. Whereas, in case of derivative causality, existing function is differentiated to deduce the other variable; and in this case no new information is added and it makes the system dependent upon already available values. Algebraically, integrals add on and produce new information whereas derivatives use the existence of the functions or variables. When an integral causality is used, it means a new state variable can be determined and when derivative causality is used, it means a state variable is dependent upon some other state variable and it is a redundant system. Physical systems can be redundant but it increases the order of the system to solve without adding more information. When assigning causality, it is preferred to avoid derivative causality in order to avoid having a redundant equation to solve; dependency on other variables as well as it may produce constraints in the model.

4.3 2-Port Elements

These elements connect two bonds together, ideally by using the law of conversation of energy for power transfer. In real scenarios, systems are not efficient to transfer complete energy into other systems without loss, but even these losses can be modeled with bond graphs by appropriately addressing the dissipation or loss. There are two basic 2-port elements, namely transformer and gyrator, ideal for power transfer linearly from one system to other.

4.3.1 *Transformer*

Ideal power transformer TF represented with two bonds in Fig. 4.3 means that power at one bond completely transfers to other without any loss by implying

$$P_1(t) = P_2(t) \rightarrow e_1(t)f_1(t) = e_2(t)f_2(t) \tag{4.15}$$

Transformer is a device that relates effort at bond one to the effort of bond two and flow of bond two with flow of bond one by a transformer ratio or modulus m.

$$\begin{aligned} e_1(t) &= m \cdot e_2(t) \\ m \cdot f_1(t) &= f_2(t) \end{aligned} \tag{4.16}$$

$$\rightharpoonup TF \rightharpoonup$$

Fig. 4.3 Bond graph of a generalized transformer

This relation can also be given as

$$\begin{aligned} n \cdot e_1(t) &= e_2(t) \\ f_1(t) &= n \cdot f_2(t) \end{aligned} \qquad n = \frac{1}{m} \tag{4.17}$$

In both equations, we can see effort at one end relates with effort at the other end of the transformer. Accordingly, flow relates with flow in both ends of the transformer, and variables m and n are transformer ratios. In order to avoid confusion and repetition, we will use only m as a transformer ratio in the text. In electrical systems, electrical transformer relates voltages and current between primary and secondary coils by turns ratio as modulus or transformer ratio. There are numerous examples of the transformer in mechanical systems, such as a lever mechanism which relates forces F_1 & F_2 at both ends and velocities V_1 & V_2 at both ends, and the transformer modulus is the ratio of lengths $m = {}^{L_1}/_{L_2}$ from the center point as shown in Fig. 4.4. The direction motion of lever will depend upon the magnitude of moment arm and accordingly the lever will move up and down from either end.

Similarly, in gears, power transmission is equal in rotation and torques τ_1 & τ_2 and angular velocities ω_1 & ω_2 are related with ratio of teeth between first gear and second gear (Fig. 4.5).

There are two choices of causality assignment for a transformer; simply if effort is known at one end of the first bond then automatically effort is automatically determined at the same end of second bond. Similarly, flows determine accordingly, one at transformer end and one away from transformer end. Causality assignments by determining flow has two choices in this case as given in Table 4.4

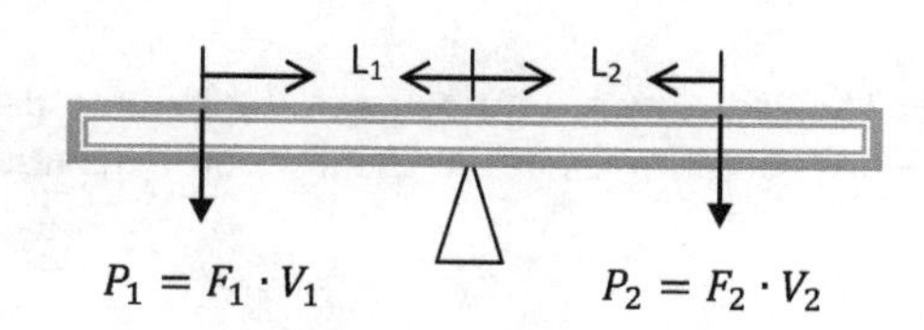

Fig. 4.4 Block diagram of a mechanical lever

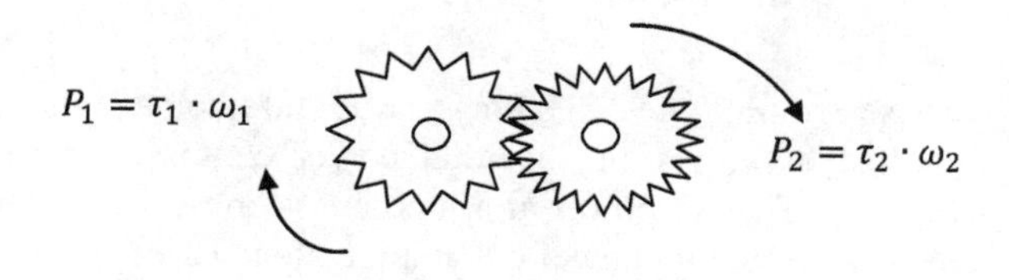

Fig. 4.5 Block diagram of a mechanical lever

Table 4.4 Bond graph of 2-port generalized transformer with constitutive law

Bond graph	Constitutive law
—1—⇀| TF —2—⇀|	$f_1(t) = \dfrac{f_2(t)}{m}$ $e_2(t) = \dfrac{e_1(t)}{m}$
|—1—⇀ TF |—2—⇀	$e_1(t) = m \cdot e_2(t)$ $f_2(t) = m \cdot f_1(t)$

$$\xrightarrow{\quad} GY \xrightarrow{\quad}$$

Fig. 4.6 Bond graph of a generalized gyrator

4.3.2 Gyrator

The second 2-port element is the gyrator represented in Fig. 4.6, which evolved from the concept of the mechanical gyroscope in which applied force affects the velocity of precession, and force in the rotational axes provides a translational motion towards the direction of force. By the law of conversation of momentum and energy, powers are equal at both sides but relating forces with velocities.

It is difficult to quote a physical device which directly acts like a gyrator in the same energy system. But there are numerous devices, sensors, and actuators which behave like gyrators, such as the dc motor which performs a gyrator action between torque at the shaft and the current of the armature and angular rate with the voltage. Another example is a speaker with a magnetic coil that has relation of sound pressure to the current in the coil and velocity with the voltage. These relationships demonstrate the use of a gyrator in the physical sense. In most of the cases, gyrator action performs the power transformation between two forms of energy. A gyrator relates the effort at first bond with the flow of second bond through gyrator ratio or modulus of gyrator r. So, it also inversely relates the flow of the first bond with the effort of the second bond.

$$\begin{aligned} e_1(t) &= r \cdot f_2(t) \\ r \cdot f_1(t) &= e_2(t) \end{aligned} \tag{4.18}$$

The causality assignments of the gyrator also have two choices, i.e., either effort or flow determines at the gyrator end as given in Table 4.5.

4.4 Junctions

These are multiport elements that require connections of three or more bonds or subsystems governing by total power conservation in all bonds represented in Fig. 4.7. The net power at a junction is conserved, which means that total power in is equal to total power out at a junction. There are two types of junctions known as 0-junction and 1-junction. These junctions represent the series or parallel connections between the subsystems. The idea of series or parallel connections is not universal when it comes to electrical or mechanical systems and also varies with generalized efforts as voltage or force, generalized flow as current or velocity. Even adding basic 1-port elements (resistor, capacitor, and inductor) in series and parallel yields different configurations; because resistor in series is the sum of all resistors, but capacitors in series are the sum of their reciprocals just like resistors in parallel.

Table 4.5 Bond graph of 2-port generalized gyrator with constitutive law

Bond graph	Constitutive law
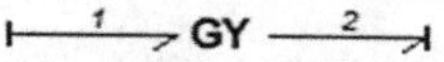 ⊢ 1 ⇀ GY — 2 ⊣⇀	$e_1(t) = r \cdot f_2(t)$ $e_2(t) = r \cdot f_1(t)$
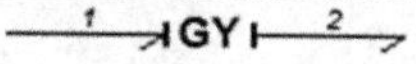 — 1 ⊣⇀ GY ⊢ 2 ⇀	$f_1(t) = \dfrac{e_2(t)}{r}$ $f_2(t) = \dfrac{e_1(t)}{r}$

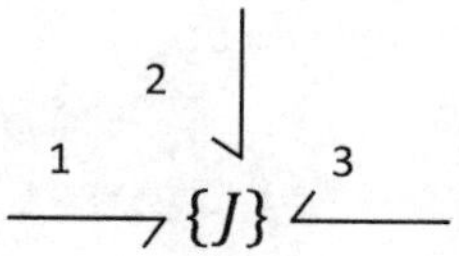

Fig. 4.7 Bond graph of a junction with three bonds, where J can be 0 or 1

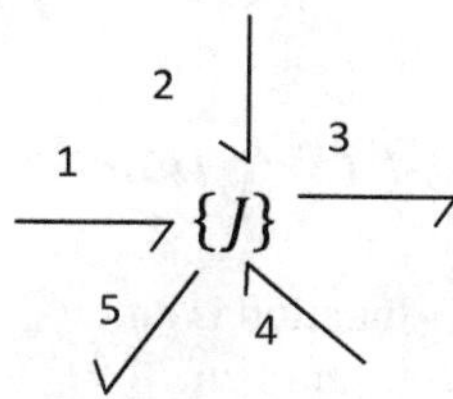

Fig. 4.8 Bond graph of a junction with five bonds, where J can be 0 or 1

Same is the case for series connection in electrical systems equivalent to the parallel connection of mechanical systems analytically. Though 0-junction and 1-junction connect subsystems in either series or parallel configurations, we cannot classify either junction as a series or parallel junction. Series and parallel connections conserve power by either fixing the effort or flow at the junction.

The power at this junction J, which is either 0-junction or 1-junction, can be written as

$$P_1(t) + P_2(t) + P_3(t) = 0 \tag{4.19a}$$

$$e_1(t)f_1(t) + e_2(t)f_2(t) + e_3(t)f_3(t) = 0 \tag{4.19b}$$

It is important to understand that this equation from bond graph depends upon direction of bond as well as causality assignment. In this bond all arrows are coming towards junction so the sum of all incoming power is equal to zero. In most cases, if some bonds are supplying power, there are other bonds that are consuming power as well so that the directions of arrows are reverse accordingly for these bonds.

Figure 4.8 shows a junction with three ports inward to the junction and two ports outward with power relation given in Eq. (4.20).

$$P_1(t) + P_2(t) + P_4(t) = P_3(t) + P_5(t) \tag{4.20}$$

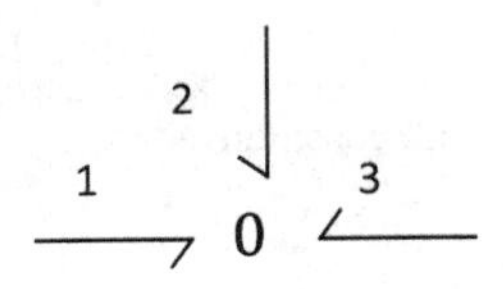

Fig. 4.9 Bond graph of a 0 junction

Table 4.6 Bond graph of a 0 junction with constitutive law

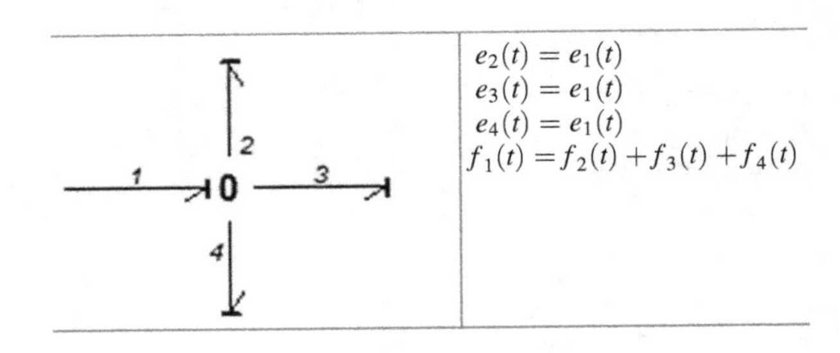

$e_2(t) = e_1(t)$
$e_3(t) = e_1(t)$
$e_4(t) = e_1(t)$
$f_1(t) = f_2(t) + f_3(t) + f_4(t)$

4.4.1 0-Junction

0-Junction is a flow junction in which flow adds at the junction while effort remains constant (Fig. 4.9).

$$e_1(t) = e_2(t) = e_3(t) \tag{4.21a}$$

$$f_1(t) + f_2(t) + f_3(t) = 0 \tag{4.21b}$$

The 0-junction represents the parallel circuits in electrical systems as a Kirchhoff's current law. In a mechanical system it represents a series connection where the same force is applied to different subsystems with different velocities. For example a bat is hitting two balls of different material at same time; the same force is applied but each ball will move with a different velocity depending upon its mass and material. There is only one possibility of causality assignment for 0-junction, i.e., a causal stroke is close to the 0-junction and all others are away from the 0-junction because it is a common effort or flow junction, which requires only one effort to be determined at the junction. This is computed by the rest of flows determined away from the junction as shown in Table 4.6. This clearly indicates that effort at bond 1 is an input to the junction and all other efforts are outputs of the junction. Conversely, flow at bonds 2, 3, and 4 are inputs and flow at bond 1 is determined algebraically at the junction. Please also note that the half arrow also indicates the same phenomenon in this case, which may not be valid in other 0-junctions with different input or output configuration.

4.4.2 1-Junction

1-Junction is an effort junction or common flow junction, where effort adds up and flow remains constant at the junction shown in Fig. 4.10. This bond represents a series combination in electrical systems using Kirchhoff's voltage law.

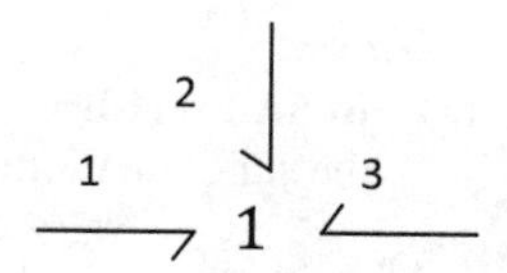

Fig. 4.10 Bond graph of a 1 junction

Table 4.7 Bond graph of a 1 junction with constitutive law

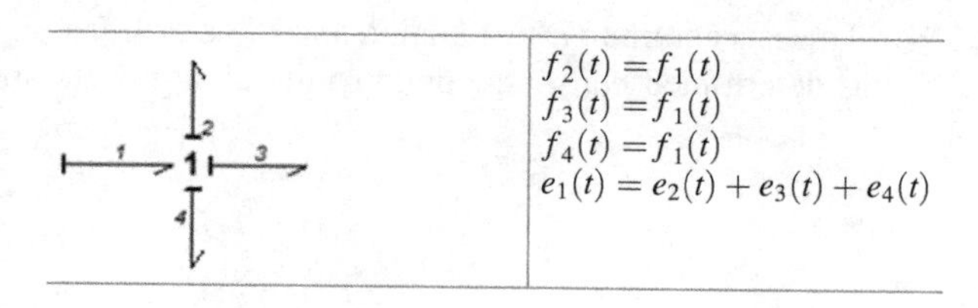	$f_2(t) = f_1(t)$ $f_3(t) = f_1(t)$ $f_4(t) = f_1(t)$ $e_1(t) = e_2(t) + e_3(t) + e_4(t)$

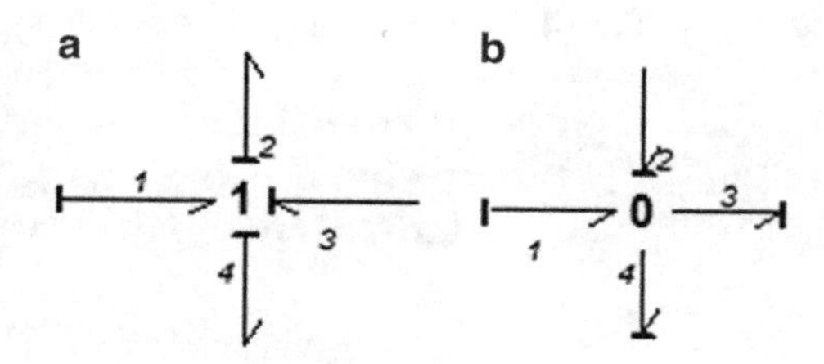

Fig. 4.11 Bond graph of 1 and 0 junction

In mechanical systems, it represents a system where different objects are moving with same velocity but with different applied forces. The constitutive law for 1-junction is given as

$$f_1(t) = f_2(t) = f_3(t) \tag{4.22a}$$

$$e_1(t) + e_2(t) + e_3(t) = 0 \tag{4.22b}$$

It is the common flow junction, so flow remains constant in all three bonds and the sum of efforts are equal to zero. This is same as sum of voltages equal to zero in a loop, or sum of forces at a point is equal to zero in translational mechanics and sum of torques equal to zero at a point in rotational mechanics. There is a single choice of causality assignment because one flow is determined at the bond and rests are equal to this flow. Effort at this bond is equal to the sum of the remaining efforts algebraically as shown in Table 4.7. In this junction, we can also observe that $f_1(t)$ is an input to the 1-junction and flows at bonds 2, 3, and 4 are equal to the $f_1(t)$. The effort of bond 1, $e_1(t)$ is determined by equating to the sum of remaining three efforts. This defining bond like bond 1 in this case is often referred as a strong bond of 1-junction and others as weak bond of this particular 1-junction. This is also clear with the half arrow direction in this specific case where power is only input from bond 1 and the distributed power is output from rest of the bonds.

Example 4.1: Equations of Junctions Write the equations of the following junctions in Fig. 4.11

Solution

(a) Bond 1 and Bond 3 are inputs to the junction and bond 2 and 4 are outputs of the junction so the equation will be

$$P_1(t) + P_3(t) = P_2(t) + P_4(t) \tag{4.23}$$

This is a common flow junction and by causality assignment we know that $f_1(t)$ is determined before the bond so remaining flows are

$$\begin{aligned} f_2(t) &= f_1(t) \\ f_3(t) &= f_1(t) \\ f_4(t) &= f_1(t) \end{aligned} \tag{4.24}$$

So Eq. (4.23) formulates as

$$e_1(t) \cdot f_1(t) + e_3(t) \cdot f_3(t) = e_2(t) \cdot f_2(t) + e_4(t) \cdot f_4(t) \tag{4.25a}$$

$$(e_1(t) + e_3(t)) \cdot f_1(t) = (e_2(t) + e_4(t)) \cdot f_1(t) \tag{4.25b}$$

$$e_1(t) + e_3(t) = e_2(t) + e_4(t) \tag{4.26}$$

Equations (4.24) and (4.26) are solution equations for this 1-junction.

(b) The 0-junction has two inputs and two outputs and this is a common effort junction. Bond 2 is a strong bond of this 0-junction and so $e_2(t)$ is determined at the junction and remaining efforts are equal to it as given in Eq. (4.27)

$$\begin{aligned} e_1(t) &= e_2(t) \\ e_3(t) &= e_2(t) \\ e_4(t) &= e_2(t) \end{aligned} \tag{4.27}$$

By equating power of the bond and factorizing common effort we directly write the equation in terms of flow variables as

$$f_2(t) = f_3(t) + f_4(t) - f_1(t) \tag{4.28}$$

4.5 Modulated Transformer and Gyrator

The modulus of a transformer and gyrator may vary depending upon other variables in the system or due to a combination of few variables. This modulus now connects through an active bond, which means it does not represent power transfer, rather it only supplies a variable criterion for transformer and gyrator ratio. The modulus is usually represented as a function of state variables in the system or may be dependent upon input or output. In electrical transformers, output voltage is usually

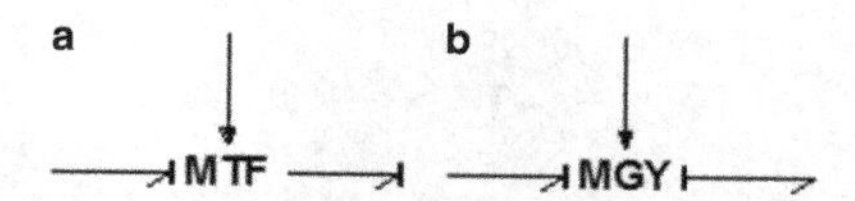

Fig. 4.12 Bond graph of modulated 2-port elements (**a**) A modulated transformer, (**b**) a modulated gyrator

fixed at a desired level and the number of turn ratio is changed depending upon load requirement and input current. A door with a hinge joint is a modulated transformer because it depends upon the angle at which force applied, and this angle and the moment arm make a torque which revolves the door. Figure 4.12 shows the graphical representation of the modulated transformer and gyrator.

Remember that there is no causality assignment for active bond graph and the modulus of transformer or gyrator represents without any causality assignment. In these cases, a modulus is a function of variable which is dependent upon for defining relationship. For example in order to relate the torque/angular rate of revolving door with applied force and translational velocity, we relate the modulus as $m(\theta)$ in a door with hinge joint. In this case θ is generalized displacement on the side of the transformer. Problem 4.5 elaborates more on this concept.

4.6 Modulated Sources

If a source represents a time function or is dependent upon some other variable, then these sources are represented as modulated sources. These sources take an input through an active bond and then supply effort or flow by a normal bond to the system. Dependent sources, like voltage-dependent voltage source, current-dependent voltage source, etc take an input from their dependencies through active bond and supplies output in the form of effort or flow to the system.

Problems

P4.1 Find the second gyrator equation and modulus r as per the definition given in Eq. (4.18) if

$$f_1(t) = s \cdot e_2(t)$$

P4.2 Find the constitutive equations at all bonds for the following two systems and explain why there is a differential causality with transformer but not with gyrator?

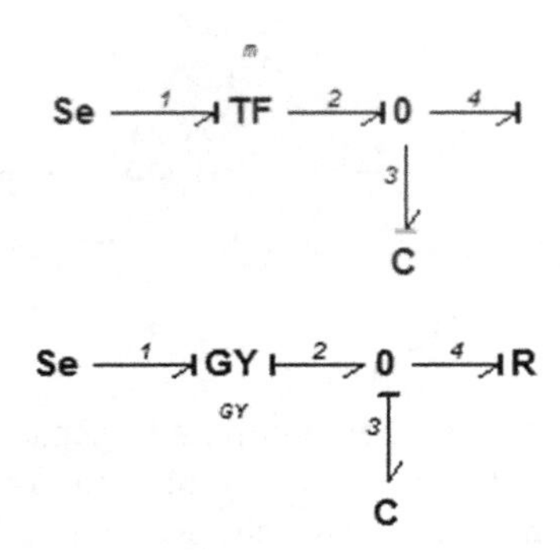

P4.3 A piston and cylinder component is shown in the figure below, which is converting mechanical translation into hydraulics system. Define a 2-port element for the system and explain its modulus with constitutive law between both systems

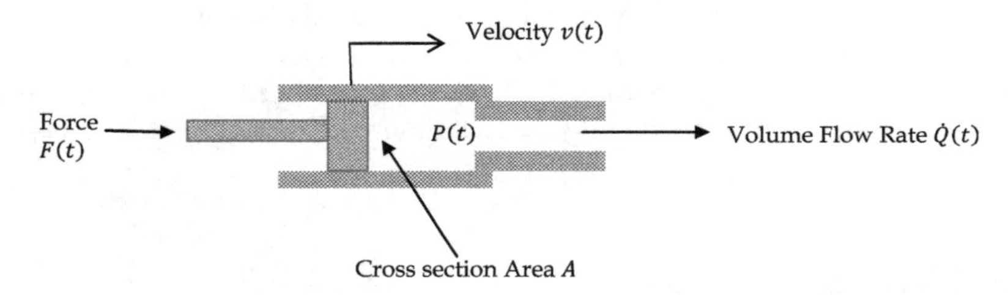

P4.4 Define constitutive laws for each element given in Tables 3.3, 3.6 and Problem 3.1 and specify the type and causality assignments for each element.

P4.5 A hinge movement of a revolute joint is shown in figure below which transforms translational energy into rotational energy. The force applied resolves into two components, and torque at hinge makes a moment arm with a component. Modulated transformer represents this scheme in bond graph methodology, and modulus is dependent upon the length of the moment arm and angle at which force is being applied. Draw a word bond graph of this scheme and formulate a constitutive with definition of modulus as the function of variables associated with the model.

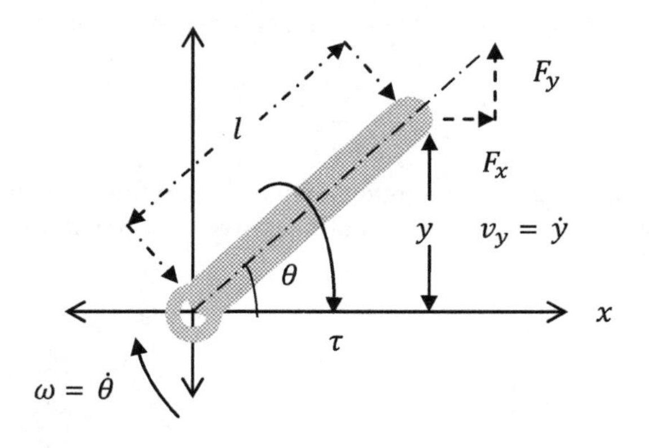

P4.6 There are modulated sources of efforts and flow which relate time as a mathematical function to effort or flow. An example is AC electrical power source with a fixed frequency component. These sources are represented as active bond graph at input with either effort or flow as a main output power source. Write the constitutive equations of following models in each subsystem with modulated sources and 2-port elements.

(a)

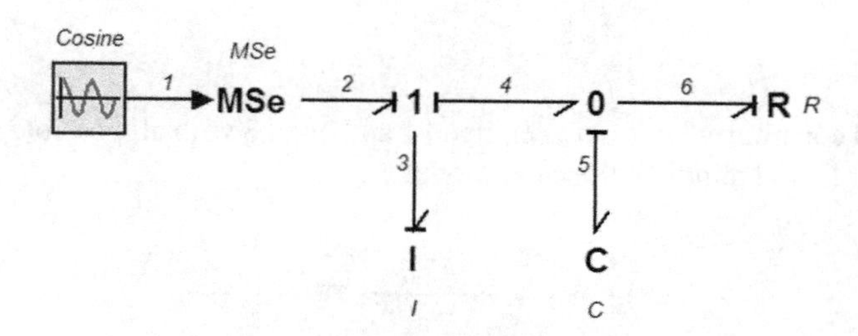

(b)

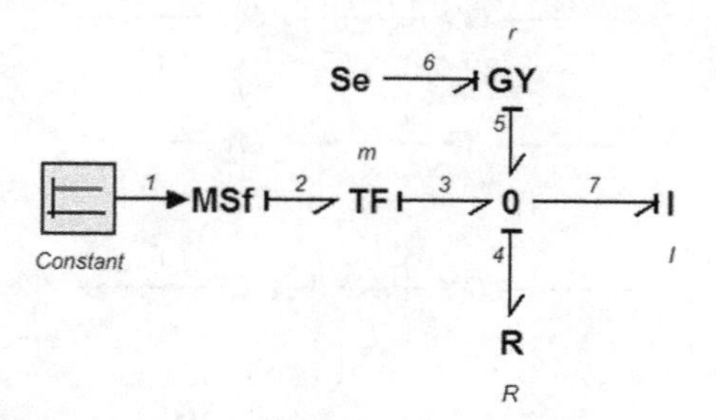

(c) Note that bond 7 is the effort of bond 6 only through active bond, represents the effort sensor in the respective bond. The bond 6 remains same before and after sensor in all respects of causality and power transfer.

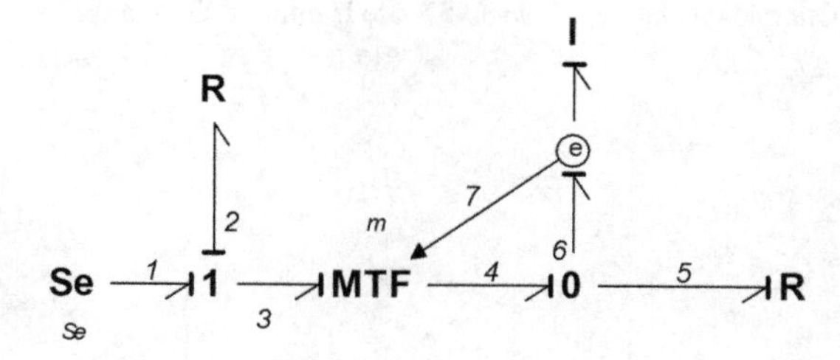

(d) Modulated modulus also defines by mathematical operation other variables as given below, which makes constitutive law in algebraic loops to write equations

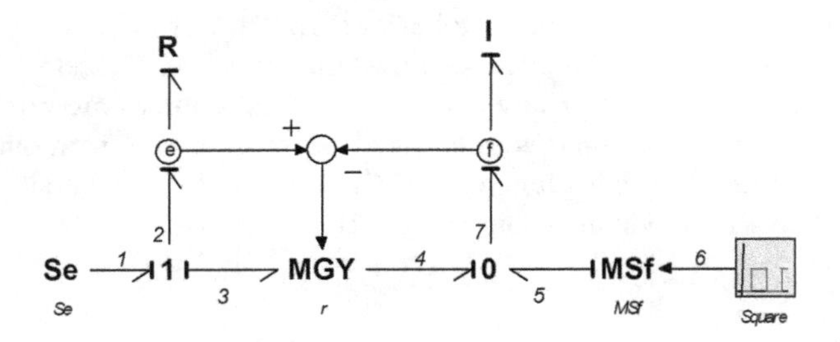

P4.7 Find a constitutive law between bond 1 and bond 3 with all possible choices of causality and simplify the each model

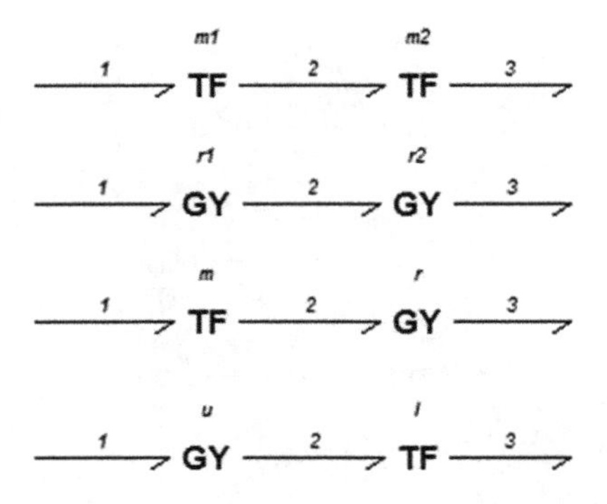

References

1. Karnopp, Dean C., Donald L. Margolis, and Ronald C. Rosenberg. 2012. *System Dynamics—Modeling and Simulation of Mechatronics Systems*, 5th ed. Hoboken: Wiley.
2. 20-Sim Software. www.20sim.com
3. François E. Cellier, and N. Ángela. 2005. The Modelica Bond Graph Library. In *Proceedings of the 4th International Modelica Conference*, 57–65. Hamburg, Germany, 7–8 Mar 2005.

Chapter 5
Analytical Formulation by Bond Graph Modeling

Bond graph modeling possesses more applicability towards systems with different energy domains. It is easier to model electrical systems with electrical theories using KCL or KVL. It is easier to model mechanical systems using Newtonian mechanics, Lagrangian mechanics, etc. Similarly, for electromagnetic systems there are Maxwell laws to deal with both electrical and magnetic forces; in hydraulics systems, there are Bernoulli's /Stroke's law to deal with pressure and forces. But now, with the advent of modern technologies, there are systems which are now amalgams of electrical, mechanical, hydraulics, electromagnetic systems, etc. Bond graph modeling provides help to these modeling schemes, and dealing with systems in their energy domains provides an easier and technically mature approach. Bond graph provides a unified approach to model a system comprised of two or more energy systems. On the other hand it is difficult to provide a universal methodology that is applicable to all systems. In electrical and hydraulic systems, nodes appear where generalized flow distributes between them whereas, in mechanical systems, nodes appear where forces act together both in series or parallel configurations. As we discussed in previous chapters, that series and parallel combination is not a universal phenomenon for modeling. So there arises a problem in formulating a unified procedure for modeling at system level. A generalized set of procedures are written for modeling the systems that is applicable for modeling systems within same energy domain as well as modeling in multi-energy domain. It needs careful attention when a series or parallel combinations of different energy systems are modeled as a single system.

5.1 Modeling Procedure

Nodes represent 0-junction or 1-junction, according to the transfer of generalized flow or effort respectively. Each 1-port and 2-port elements are attached with additional 0/1 junction between nodes and then bond graph is simplified to get

A.M. Mughal, *Real Time Modeling, Simulation and Control of Dynamical Systems*,
DOI 10.1007/978-3-319-33906-1_5

system equations. A step-by-step construction procedure formulates block diagram to the bond graph representation. It is better to make a comprehensive block diagram of the system and then make the word bond graph if required and especially in case of power transfer between energy systems. A unified bond graph construction procedure is given as:

Step 1: Identify nodes where either effort or flow distributes in the system.
Step 2: Place 0-junction at every node where effort is same and flow is distributing, if there is no junction where 0-junction can be placed then go to Step 3.
Step 3: Place 1-junction at every node where flow is the same and the effort is distributing, ensure that both 0-junction and 1-junction cannot be placed at the same node.
Step 4: Insert 0- and 1-junction between 0 and 1 at nodes such that 0 and 1 alternate each other in a sequence.
Step 5: Add 1-port & 2-port elements to 0/1 junctions inserted where applicable.
Step 6: Add half-arrow bonds between 0-junctions, 1-junctions, and 1-port & 2-port elements such that the direction of the arrow is from sources to the 1-port elements.
Step 7: Delete at least one ground or reference junction, such as a ground node in electrical system, reference velocity node, (reference pressure node in hydraulics). If a ground or reference junction doesn't appear then it may be avoided, it usually occurs when different energy systems model together in a loop. It may also occur if a block diagram or word bond graph does not assign any node to ground or reference junction.
Step 8: Simplify the bond graph by deleting extra bonds such that 0 and 1 alternates; all junctions must have three or more bonds.
Step 9: Verify the flow of power after simplification and basic criteria should not be disturbed for distributing effort or flow as in Step 1.
Step 10: Assign causality in order, first to power sources, second to capacitors and inductors for integral causality, then 2-port elements and resistors by not violating the required causality of junctions {Remember the choices of causalities for resistors, capacitors, inductors, transformers, and gyrators—preference should be given to integral causality but differential causality can be used if deemed necessary}.
Step 11: Assign number to the bonds and get ready to write state space equations of the system.

Example 5.1 A simple electrical circuit (Fig. 5.1)

Now we can follow the step-by-step procedure to generate a bond graph of the model.

Step 1: From the figure, it shows six nodes between components but there is only one node at ground, V_{DC}, R and I (inductor) connect at the same ground. Efforts are distributing between V_{DC} and C, R_2 and I, and between C and R_1 in parallel with R_2 and I.
Step 2: We place 0 s at the nodes where efforts are different as (Fig. 5.2):

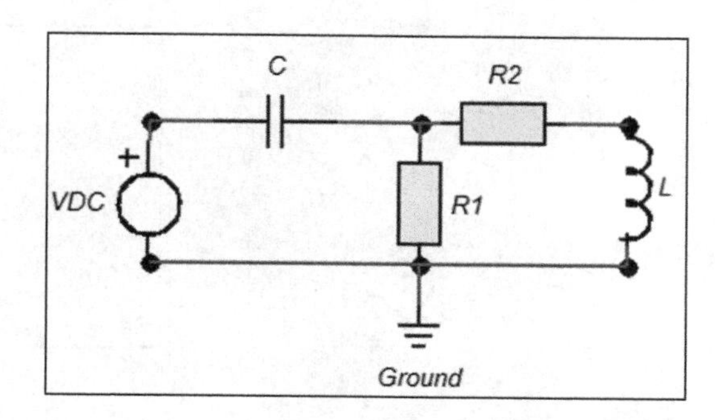

Fig. 5.1 Block diagram of an electrical circuit

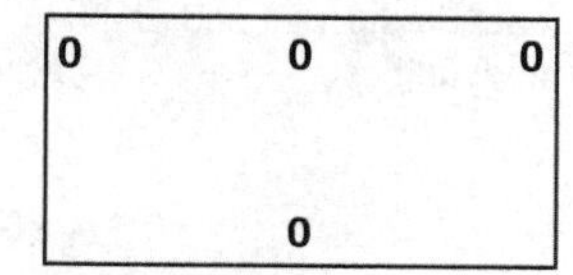

Fig. 5.2 Step 2 of procedure for block diagram in Fig. 5.1

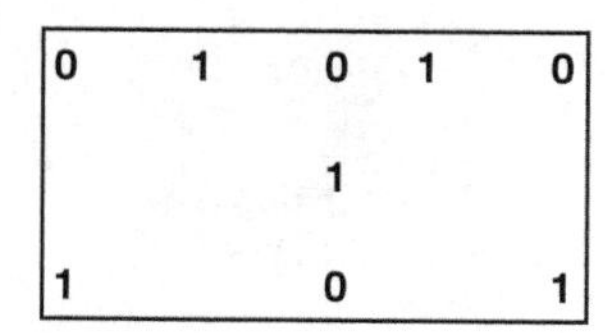

Fig. 5.3 Step 4 of procedure for block diagram in Fig. 5.1

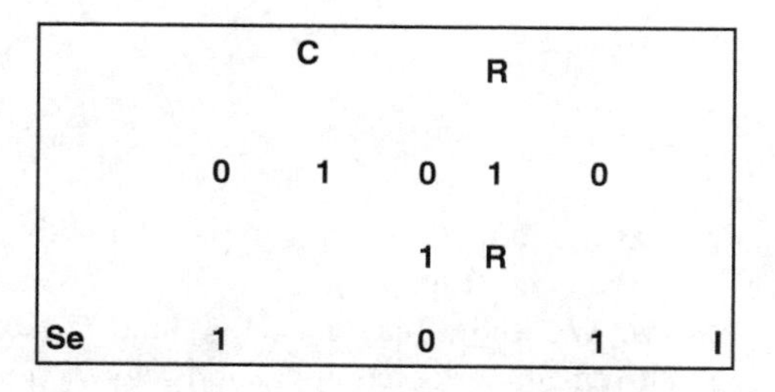

Fig. 5.4 Step 5 of procedure for block diagram in Fig. 5.1

Step 3: There are nodes where flow is different but a zero is already placed, so skip to next step.

Step 4: Inserting 0 and 1 junction between 0 and 1 such that 0 and 1 alternate (Fig. 5.3).

Step 5: Insert elements (Fig. 5.4).

Step 6: Insert bonds (Fig. 5.5).

Note: Elements are joined with inserted 0/1 junctions between the bonds.

Step 7: Delete reference node (Fig. 5.6).

Step 8: Deleting extra bonds and simplifying: There are two 0-junctions with less than three ports and a 1-junction with less than three ports. 1-0-1 can be simplified as 1 because both S_e and C have the same flow, similarly R_2 and

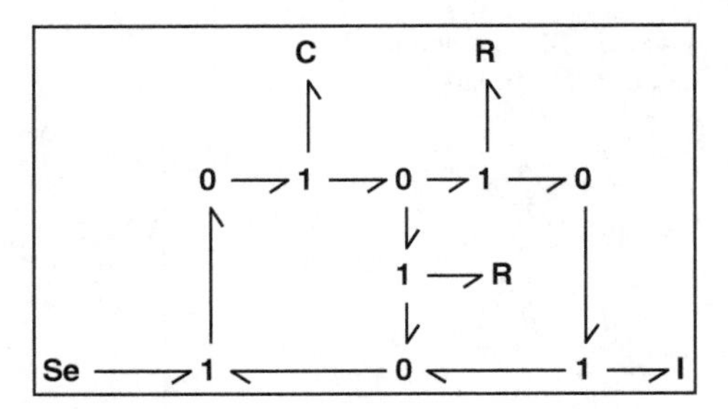

Fig. 5.5 Step 5 of procedure for block diagram in Fig. 5.1

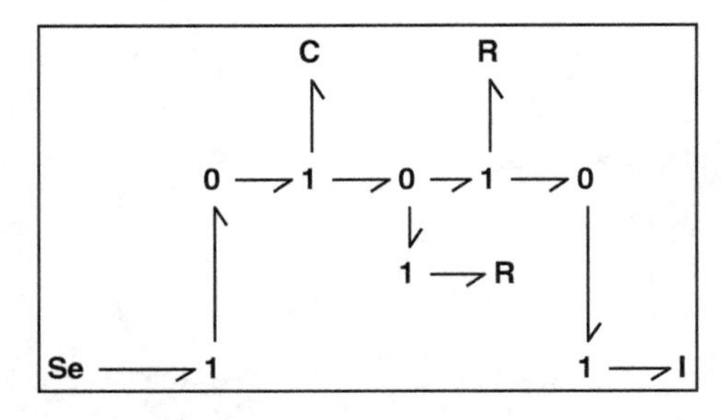

Fig. 5.6 Deleting a ref junction during bond graph modeling

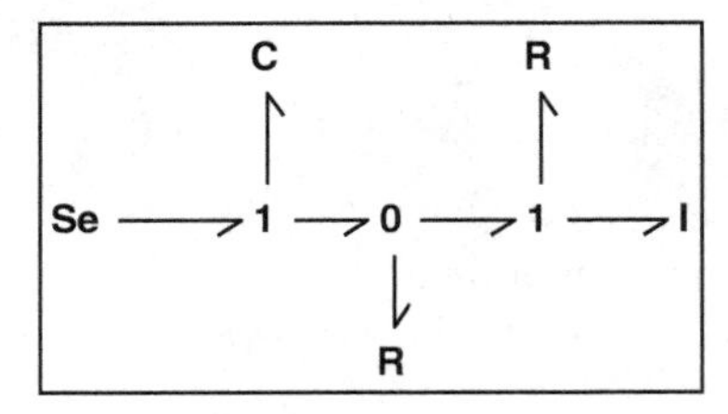

Fig. 5.7 Assigning power directions of the bond

I has the same flow. R_1 is in parallel with two 1-junctions, so this 1-junction can be removed (Fig. 5.7).

Step 9: We know that $S_e = V_{DC}$ and C have the same flow but different effort (current in series components) and R_2 and I have the same flow and different effort. Flow is distributing between three links in series represented by a 0-junction, where effort is the same in these links. This combination logically conforms to the series and parallel connections in the circuit and verifies the bond graph representation.

Step 10: Causality assignment is easy in this case, as we assign causality to source of effort S_e and integral causality to C and I, the rest will follow in line.

Step 11: Finally numbers are assigned to each bond in the model (Fig. 5.8).

Example 5.2 Modified Circuit (Fig. 5.9)

If we compare with the previous example, we can directly draw bond graph till Step 9 as follows (Fig. 5.10):

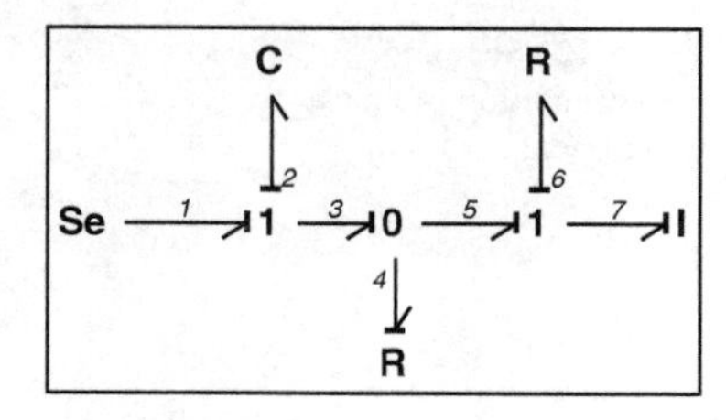

Fig. 5.8 Final bond graph of an electrical circuit given in Fig. 5.1

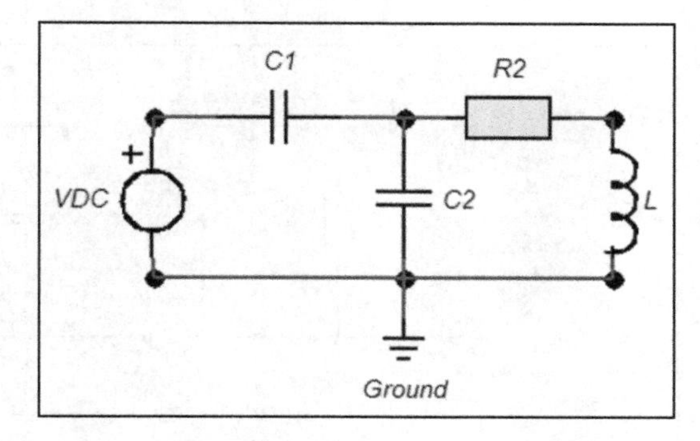

Fig. 5.9 Block diagram of a modified electrical circuit

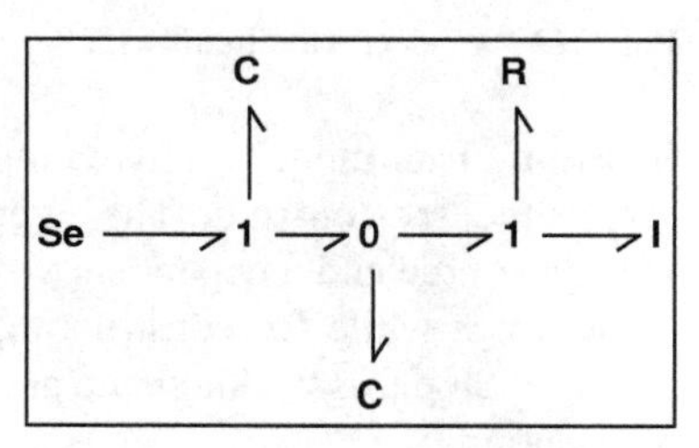

Fig. 5.10 Step 2 of procedure for block diagram in Fig. 5.9

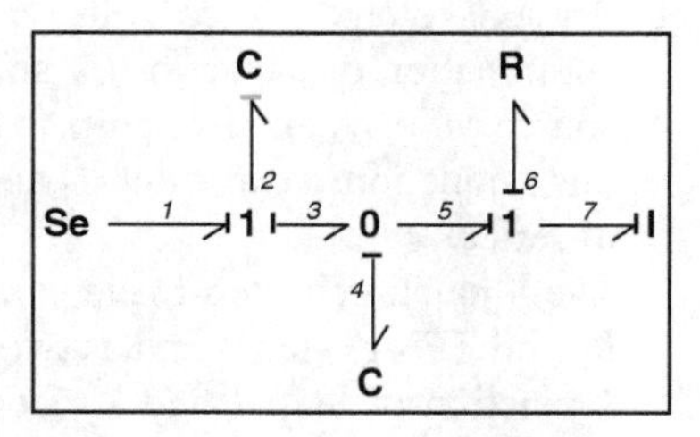

Fig. 5.11 First Representation of causality assignment

Step 10/11: There are two choices of causality assignments depending upon which capacitor is chosen for differential causality (Figs. 5.11 and 5.12).

Example 5.3 Dependent Sources

Draw a bond graph of the following system with a voltage-controlled current source for measuring current passing through resistor R_6 and voltage drop across R_2 (Fig. 5.13).

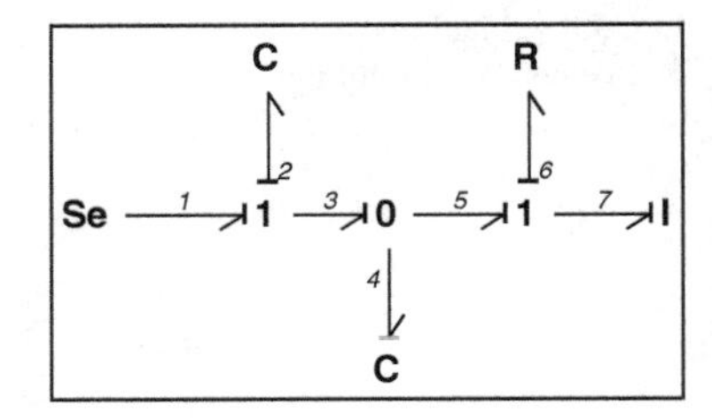

Fig. 5.12 Second Representation of causality assignment

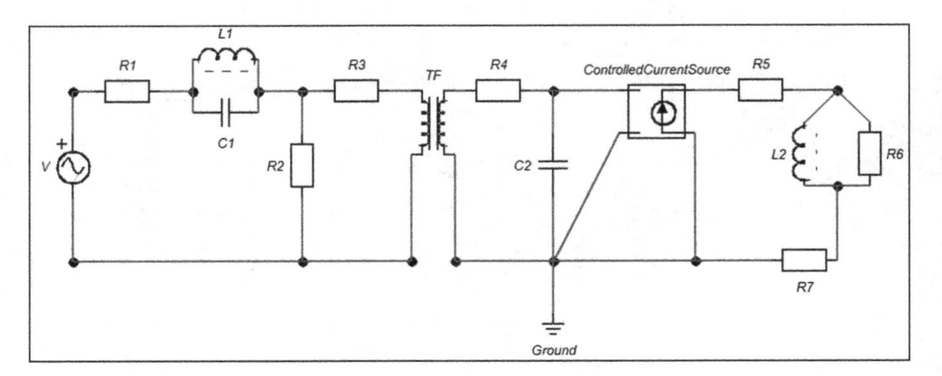

Fig. 5.13 An electrical circuit with dependent sources

Solution This model seems complicated but in reality it is a very simple bond graph model to construct. The complete system is an electrical network with an AC voltage source and a dependent voltage control current source. We leave measurements for a while for constructing a bond graph representation. We can directly reach till Step 8 as we know the points where effort is distributing and where flow is distributing.

1. An AC voltage source is in series with resistor R_1 and in series with parallel combination of capacitor C_1 and inductor L_1. A 1-junction connects a voltage source MSe, R_1 and two ports to 0-junctions in this part. A 0-junction connecting this 1-junction, a capacitor C_1 and Inductor L_1 represent the parallel combination in series.
2. The 1-junction in step 1 is in parallel with resistor R_2 and a series combination of R_3 and TF. A 0-junction has two 1-junctions on both sides; one connecting with 1-junction of Step 1 and a second connecting with a series combination of R_3 and TF.
3. The other side of the transformer is also connected with a 1-junction, which connects R_4 and a 0-junction.
4. The 1-junction of Step 3 is followed by a 0-junction, which connects a capacitor C_2 and a voltage-controlled current source MSf.
5. The output of modulated source of flow MSf is connected with a series combination of R_5, R_7, and a parallel combination of L_2 and R_6 represented by 1-junction connecting a 0-junction like Step 1.

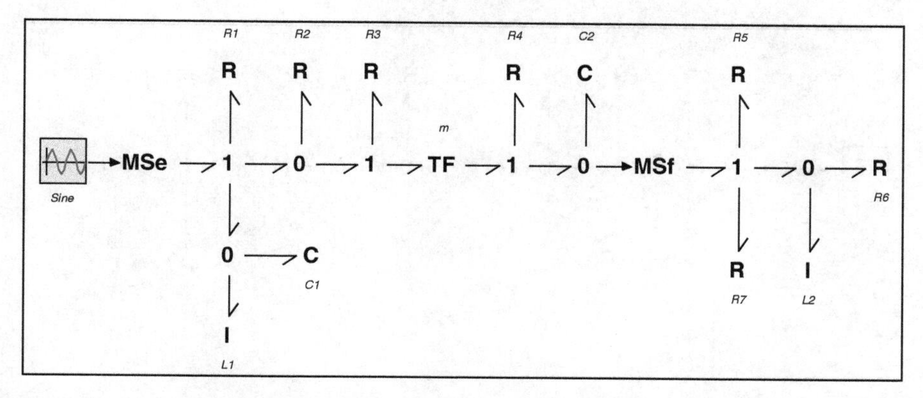

Fig. 5.14 Bond graph of an electrical circuit given in Fig. 5.13

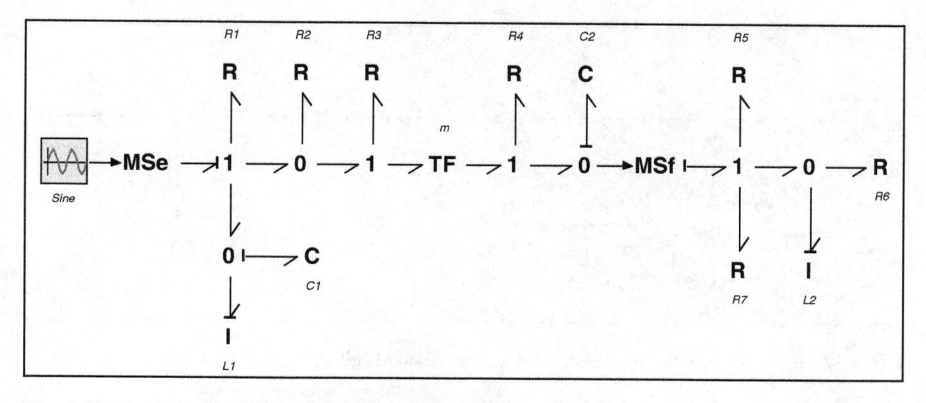

Fig. 5.15 Causality assignments for bond graph of Fig. 5.13

The bond graph of the system is represented as follows (Fig. 5.14).

The logic can be verified by series and parallel combination for effort and flow junctions. There is no need to delete any ground as during this process ground junction does not appear in the loop. Resistors R_5 and R_7 are in series and can be evaluated as R_{eq} but if in process effort across both is desirable then separate resistors (components) should be placed. Assigning causality is also simple if we follow from causality of sources (*MSE* and *MSf*), integral causality for all capacitors and resistors (Fig. 5.15).

The causalities of C_1 and C_2 force the single choice for transformer causality after assigning causalities to respective 0-junctions and 1-junctions. At the end, causalities of resistors are assigned thus completing the causality assignment followed by numbering of bonds (Fig. 5.16).

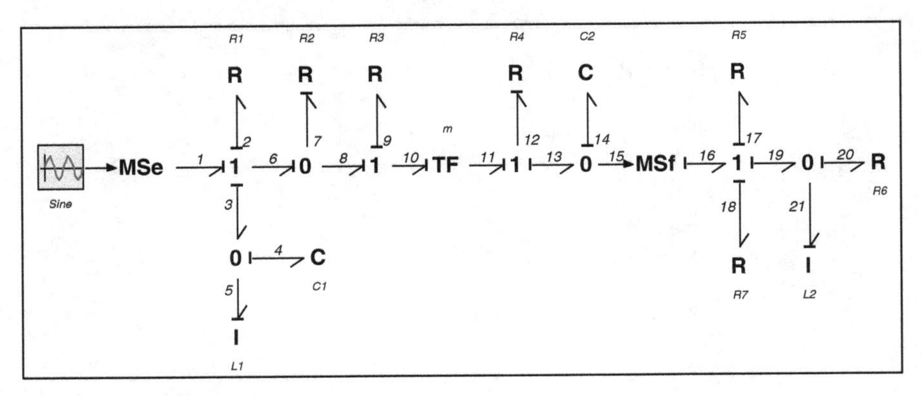

Fig. 5.16 A complete bond graph of an electrical network given in Fig. 5.13

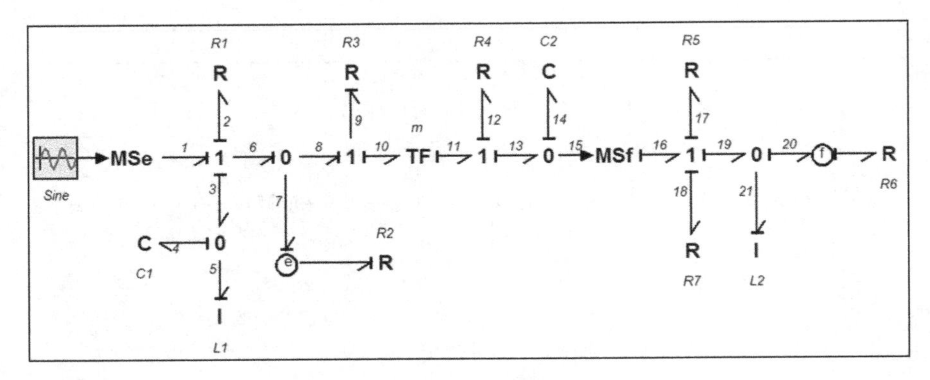

Fig. 5.17 A complete bond graph of an electrical network given in Fig. 5.13 with two sensors

5.2 Sensors

The current passing through R_6 and voltage drop in R_2 can be measured by placing sensors in 20-Sim software and these constitute the required measured output vector. A sensor for each requirement can be placed using 20-Sim software, and there is no need to assign a new bond number because it does not change causality or mathematical formulation. These sensors are used for simulation purposes for monitoring desired response or feedback to the control mechanism (Fig. 5.17).

5.3 Mechanical Systems

Example 5.4 Spring, Mass, & Damper System

Solve the classical mechanical system in spring, mass, and damper configuration (Fig. 5.18).

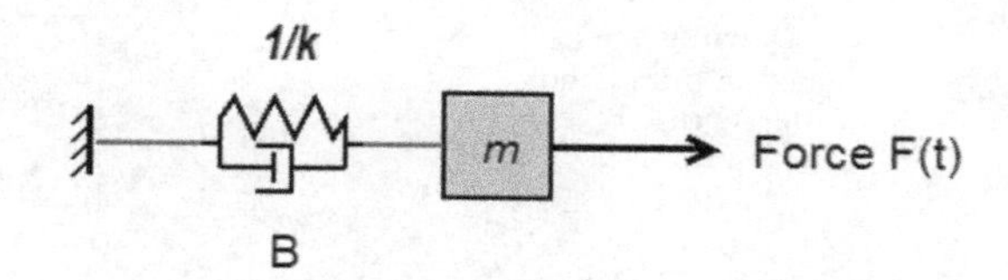

Fig. 5.18 A simple mass-spring-damper system

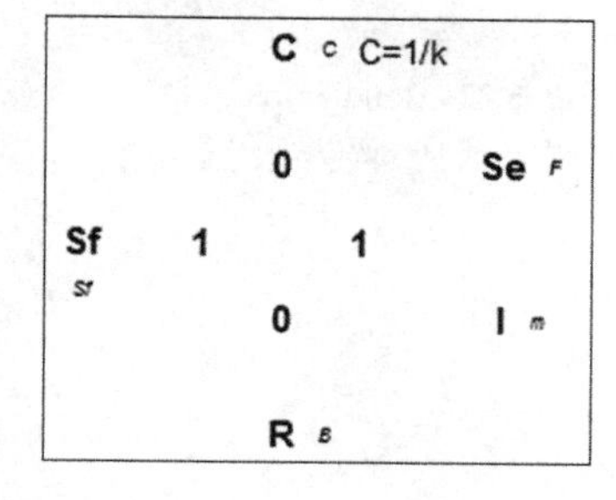

Fig. 5.19 Placing 0/1 junction and respective element for system in Fig. 5.18

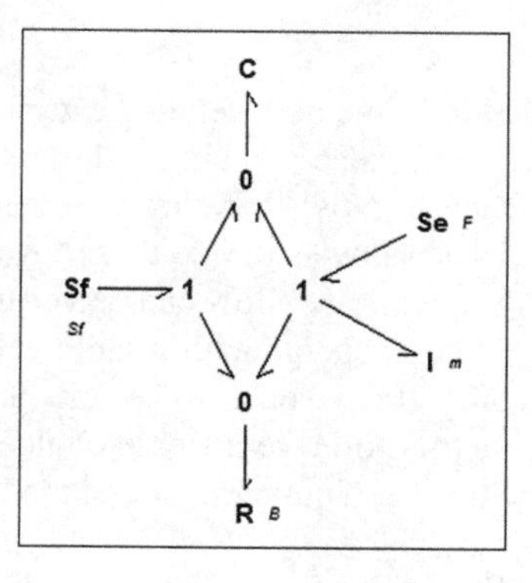

Fig. 5.20 Interconnection of half-arrow bonds from sources

Solution The difference between mechanical and electrical system is the logic of series and parallel combination. In mechanical systems, velocities are measured with respect to ground or reference. 1-junctions are placed for different velocity nodes in the system.

Step 1: There are three nodes in this system with different velocities, zero velocity at ground, velocity between spring, mass, & damper and at the other end of mass where force is pulling the mass.

Step 2: In this system there is one uniquely determined effort so we skip to Step 3 and flow distribution varies with different nodes.

Step 3: There are two places where effort is distributing and the flow is same, so we place 1-junctions at two points.

Step 4: Insert zeros between 1-junctions

Step 5: Add 1-port elements to the zeros (Fig. 5.19)

Step 6: Add half-arrow bonds between the junctions (Fig. 5.20)

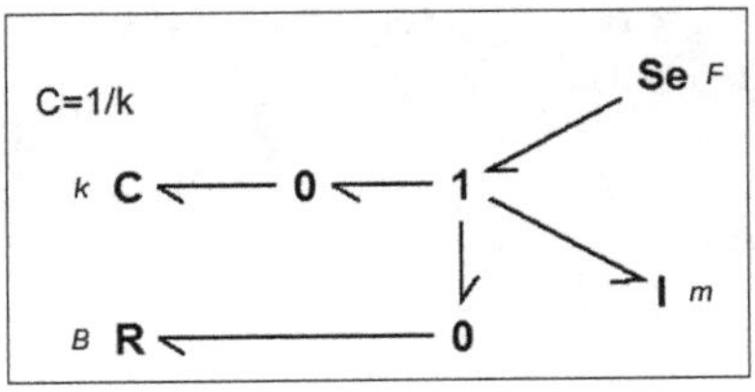

Fig. 5.21 Deleting a ground junction from bond graph of Fig. 5.20

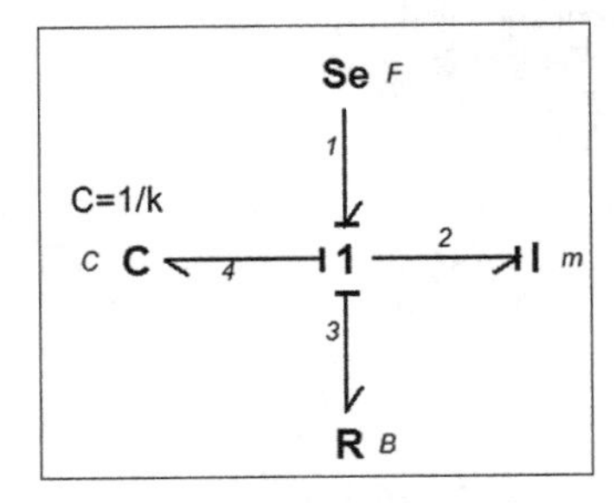

Fig. 5.22 Bond graph of a simple mass-spring-damper system

Step 7: We now delete the ground represented as the source of flow which is zero or reference velocity in the system (Fig. 5.21).
Step 8: Now there are two zero junctions and a 1-junction which have only two ports, we squeeze these together.
Step 9: Power flow can be verified that a single source of effort is pulling with three components with a same velocity and different distribution of applied force.
Step 10: A source of effort, integral casualties for capacitor (spring), and inertia (mass) leave a single choice for resistor (damper).
Step 11: Numbers are assigned for four bonds in the system (Fig. 5.22).

Example 5.5 An Electromechanical Network

A closed loop controlled system is used to adjust the position of a mass m hanging vertically downward through a complex assembly. A DC motor connected with shaft of inertia J is controlling a crank and rod assembly by adjusting crank angle a to fix the rod connected with spring (k). This spring-connected spring adjusts the rack and pinion by rotating a wheel of radius r. This movement causes a movement in upward or downward direction through a pulley. The relative position of mass is fed back via position sensor to the modulated power source which controls the DC motor (Fig. 5.23).

Solution Step 1/Step 2/Step 3:

1. The first portion is an electrical system with inductor and resistor connected in series with modulated voltage source and a DC motor. These four components are joined as a 1-junction.
2. The DC motor is connected with a shaft of inertia J. Motor is represented by a Gyrator supplying electrical power to mechanical power through a modulus μ.

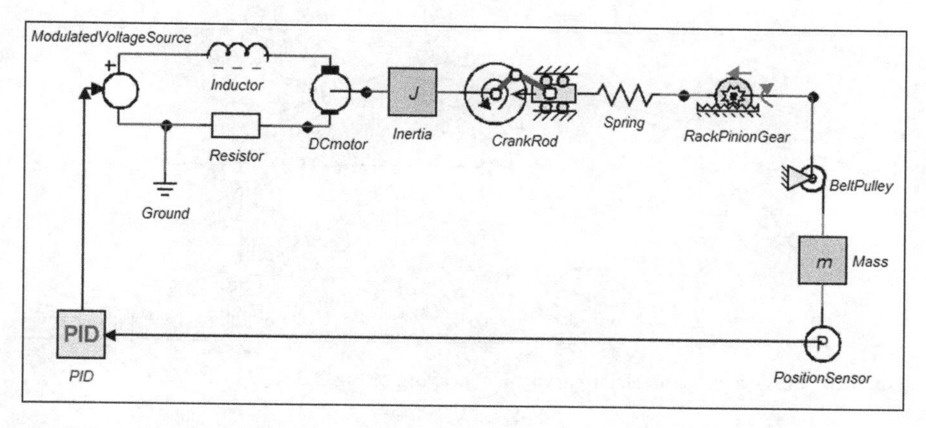

Fig. 5.23 An electromechanical network

The other end is also a 1-junction, which is connected with gyrator, inertia, and a transformer.

3. The crank rod mechanism is represented as a transformer as it relates the torque to translational force as well as angular velocity of wheel to translational velocity of rod.
4. The other end of this transformer is connected with a spring, this is represented by a 0-junction where force remains the same for a rod pulling the spring followed by a spring pulling a belt.
5. Belt pulley is directly connected to the mass so it is not transforming variable so there is no need to connect any element to a fixed belt pulley.
6. The transformer of Rack-Pinion system is now connected to a 1-junction where speed or mass remains the same under two different forces. One force is applied through DC motor via mechanical assembly and the other is gravitational force.
7. A position sensor is a generalized displacement sensor connecting to PID control system which provides feedback to modulated source.

Step 4: There is no need to place 0 or 1 junction in between existing junctions.

Step 5: There are five elements that are directly connected to the respective 0 and 1-junctions. One modulated source of effort and force under gravity connects with the first and last 1-junctions respectively.

Step 6: Half-arrow direction is from modulated source of effort to the rest of components except only for a second source $F = mg$ which also directs power from source to mass m.

Step 7: There is no ground or reference point in electromechanical loop.

Step 8: Already simplified

Step 9: Flow of power is from modulated source of effort to rest of the system.

Step 10: First we assign causality to both source of efforts, integral causality to three generalized inertia element and a generalized capacitive element. Starting for 1-junction from either end, the causality of a gyrator and both transformers

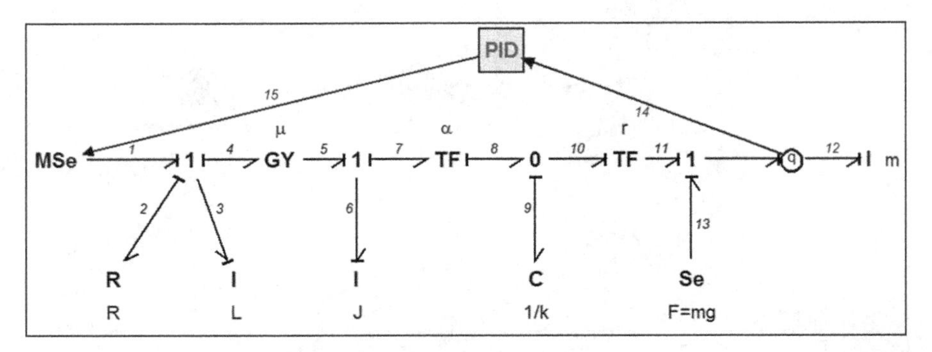

Fig. 5.24 Bond graph of an electromechanical network of Fig. 5.23

are uniquely determined by following laws at 0 and 1-junctions. We can observe a single choice of causality of the whole system without any differential causality.

Step 11: Numbering is then assigned from the modulated source to the mass under observation. Figure 5.24 shows the complete bond graph of the system.

5.4 State Space Equations

State space equations from bond graph models can be directly obtained in matrix form by writing algebraic equations at each 0-junction and 1-junction. In order to start writing state space equation, we must define state and input vectors. All sources constitute the input vector in any order of choice. State vector is defined by energy storing elements i.e., generalized capacitance and generalized inertia. *Generalized displacement of each capacitor and generalized momentum of each inertia element define the state variables*. The differential of generalized displacement is flow in the capacitor and differential of generalized momentum across the inertia is effort. Algebraic solution of efforts at 1-junction and flow at 0-junctions leads to state space formulation of the model.

Example 5.6 Equation Derivations-1

Write state space equations of the following bond graph (Fig. 5.25)

Solution In this bond graph we know that there is only one source and two energy storing elements. So we define the state vector and input vector as

$$\vec{x}(t) = \begin{bmatrix} q_2(t) \\ p_7(t) \end{bmatrix}, \quad \vec{u}(t) = [S_e] \tag{5.1}$$

Generalized displacement in 1-port capacitor is denoted by $q_2(t)$, where q is a variable for generalized displacement and subscript 2 represents that the variable

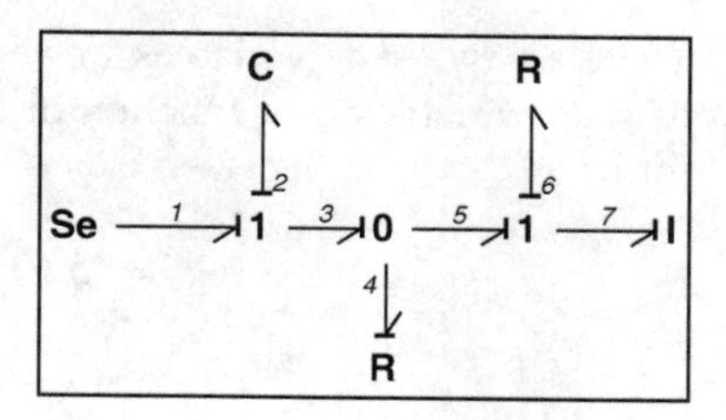

Fig. 5.25 A bond graph representation with a capacitor and an inductor

associates with bond number 2. Similarly $p_7(t)$ represents the generalized momentum related with bond 7 of 1-port inertia. Source of effort S_e constitutes an input vector of the system and in this case there is no need to represent numbers with it. Now we can write algebraic equations for each 1-junction and 0-junction of the model. 1-junction is a common flow and 0-junction is a common effort, so the three set of equations are

1-junction

$$e_3(t) = e_1(t) - e_2(t) \tag{5.2a}$$

$$f_3(t) = f_1(t) = f_2(t) \tag{5.2b}$$

0-junction

$$e_3(t) = e_4(t) = e_5(t) \tag{5.3a}$$

$$f_3(t) = f_4(t) + f_5(t) \tag{5.3b}$$

1-junction

$$e_7(t) = e_5(t) - e_6(t) \tag{5.4a}$$

$$f_7(t) = f_6(t) = f_5(t) \tag{5.4b}$$

Now we write equations in terms of state variable and inputs

$$e_1(t) = S_e \tag{5.5}$$

$$e_2(t) = \frac{q_2(t)}{C} \tag{5.6}$$

$$f_2(t) = \dot{q}_2(t) \tag{5.7}$$

$$e_7(t) = \dot{p}_7(t) \tag{5.8}$$

$$f_7(t) = \frac{p_7(t)}{I} \tag{5.9}$$

There are two state variables so we should get two equations involving differential of state variables as a function of state variables and input. Equation (5.3b) now forms as

$$\dot{q}_2(t) = f_4(t) + f_7(t) \tag{5.10}$$

Whereas

$$f_4(t) = \frac{e_4(t)}{R} \tag{5.11}$$

$$e_4(t) = e_3(t) = S_e - \frac{q_2(t)}{C} \tag{5.12}$$

So

$$\dot{q}_2(t) = \frac{1}{R}\left(S_e - \frac{q_2(t)}{C}\right) + \frac{p_7(t)}{I} \tag{5.13}$$

Now we write Eq. (5.4a) as

$$\dot{p}_7(t) = e_5(t) - e_6(t) \tag{5.14}$$

$$\dot{p}_7(t) = e_3(t) - R \cdot f_6(t) \tag{5.15}$$

$$\dot{p}_7(t) = S_e - \frac{q_2(t)}{C} - R \cdot \frac{p_7(t)}{I} \tag{5.16}$$

These two equations represent a state space model given as

$$\begin{bmatrix} \dot{q}_2(t) \\ \dot{p}_7(t) \end{bmatrix} = \begin{bmatrix} -\frac{1}{RC} & \frac{1}{I} \\ -\frac{1}{C} & -\frac{R}{I} \end{bmatrix} \cdot \begin{bmatrix} q_2(t) \\ p_7(t) \end{bmatrix} + \begin{bmatrix} \frac{1}{R} \\ 1 \end{bmatrix} \cdot [S_e] \tag{5.17}$$

Example 5.6 Equation Derivations-2

Find a relationship to determine the force exerted on a spring in Example 5.4 (Fig. 5.26)

Solution In this example there are also two state variables and state/input vectors can be written as

$$\vec{x}(t) = \begin{bmatrix} p_2(t) \\ q_4(t) \end{bmatrix}, \quad \vec{u}(t) = [S_e] = [F(t)] \tag{5.18}$$

Fig. 5.26 Bond graph of a simple mass-spring-damper system

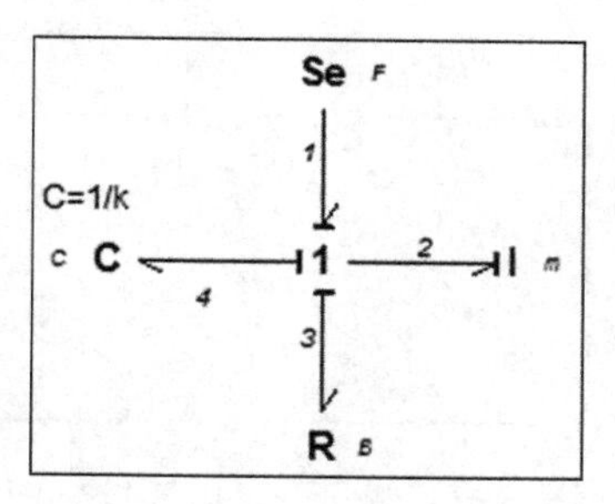

It is important to note that $C = 1/k$, which means that generalized capacitance is reciprocal of stiffness of the spring. As there is only one 1-junction, so we can write an equation for effort junction as

$$e_1(t) = e_2(t) + e_3(t) + e_4(t) \tag{5.19}$$

Where

$$e_2(t) = \dot{p}_2(t) \tag{5.20}$$

$$e_4(t) = k \cdot q_4(t) \tag{5.21}$$

$$e_3(t) = f_3(t) \cdot B = f_2(t) \cdot B = \frac{p_2(t)}{m} \cdot B \tag{5.22}$$

So Eq. (5.19) can be written as

$$\dot{p}_2(t) = F(t) - k \cdot q_4(t) - \frac{p_2(t)}{m} \cdot B \tag{5.23}$$

The other equation is a flow equation of the junction as $f_4(t) = f_2(t)$

$$\dot{q}_4(t) = \frac{p_2(t)}{m} \tag{5.24}$$

The state space formulation for measuring the force is given as

$$\begin{bmatrix} \dot{p}_2(t) \\ \dot{q}_4(t) \end{bmatrix} = \begin{bmatrix} -\frac{B}{m} & -k \\ \frac{1}{m} & 0 \end{bmatrix} \cdot \begin{bmatrix} p_2(t) \\ q_4(t) \end{bmatrix} + \begin{bmatrix} 1 \\ 0 \end{bmatrix} \cdot [F(t)]$$
$$\vec{y}(t) = e_4(t) = [0 \quad k] \cdot \begin{bmatrix} p_2(t) \\ q_4(t) \end{bmatrix} \tag{5.25}$$

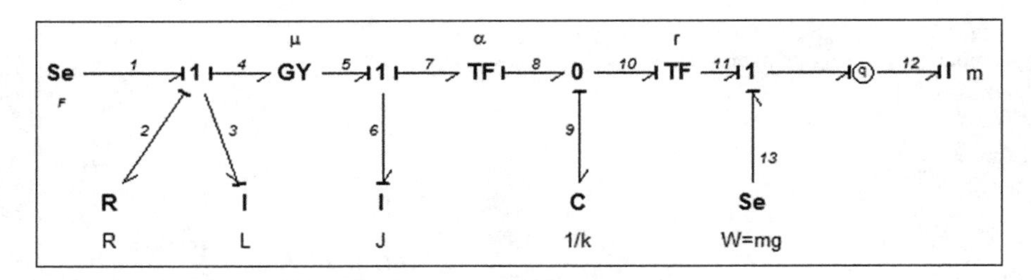

Fig. 5.27 Bond graph of an electromechanical network of Example 5.5 without controller

Example 5.7 Equation Derivations-3

Formulate a state space system for Example 5.5 in order to design a controller to regulate the position and velocity of a mass m by a DC voltage source.

Solution The bond graph of the system does not include a feedback loop for control and a simple bond graph representation is as follows (Fig. 5.27):

This is the same bond graph given in Fig. 5.24 with some differences: a PID controller and two feedback bonds are eliminated, and modulated source of effort is replaced with simple source of effort S_e. There are three inductors and a capacitor to define a state vector and two sources of efforts to define an input vector given as

$$\vec{x}(t) = \begin{bmatrix} p_3(t) \\ p_6(t) \\ q_9(t) \\ p_{12}(t) \end{bmatrix}, \quad \vec{u}(t) = \begin{bmatrix} F(t) \\ W \end{bmatrix} \tag{5.26}$$

There are two transformers and a gyrator with the following governing equation as

$$\begin{aligned} e_4(t) &= \mu \cdot f_5(t) \\ e_5(t) &= \mu \cdot f_4(t) \end{aligned} \tag{5.27}$$

$$\begin{aligned} f_8(t) &= \alpha \cdot f_7(t) \\ e_7(t) &= \alpha \cdot e_8(t) \end{aligned} \tag{5.28}$$

$$\begin{aligned} e_{11}(t) &= \frac{e_{10}(t)}{r} \\ f_{10}(t) &= \frac{f_{11}(t)}{r} \end{aligned} \tag{5.29}$$

Equations at first 1-junction are written as

$$e_1(t) = e_2(t) + e_3(t) + e_4(t), \tag{5.30}$$

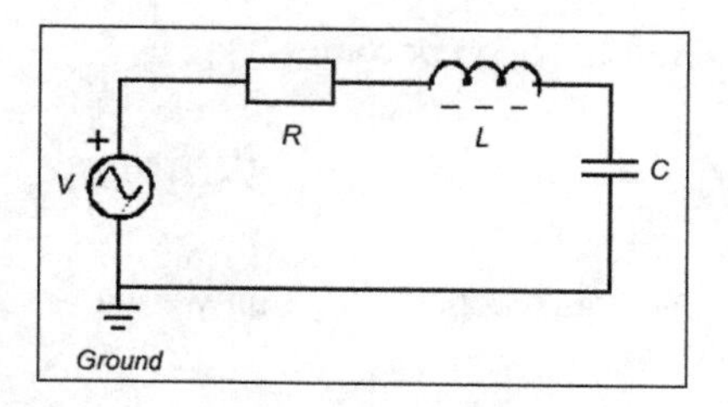

Fig. 5.28 A filter using RLC components

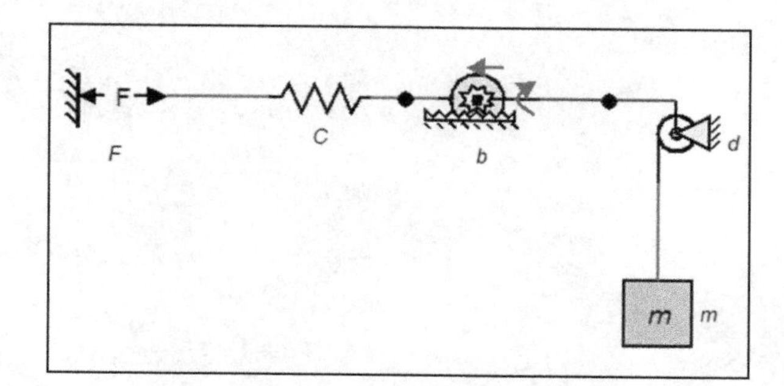

Fig. 5.29 A spring and mass system with two rack-pinion and pulley

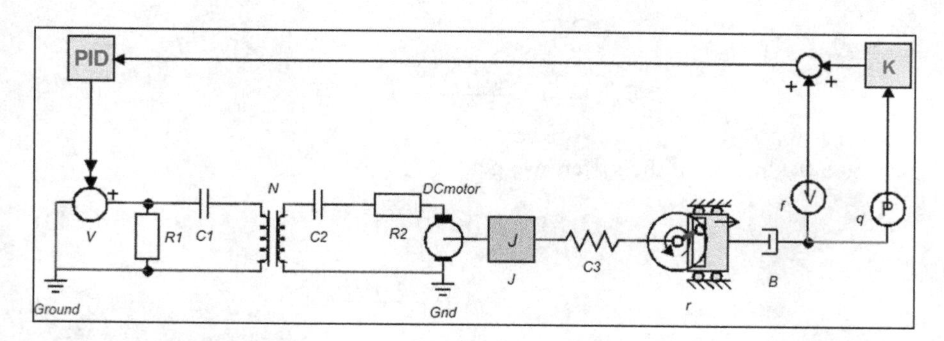

Fig. 5.30 A hybrid electromechanical network with control system in the feedback loop

Now replacing with state and input variables yields these equations as

$$\dot{p}_3(t) = F(t) - R \cdot f_2(t) - \mu \cdot f_5(t) \tag{5.31}$$

Where

$$f_2(t) = f_3(t) = \frac{p_3(t)}{L} \tag{5.32}$$

and

$$f_5(t) = f_6(t) = \frac{p_6(t)}{J} \tag{5.33}$$

So Eq. (5.31) becomes

$$\dot{p}_3(t) = F(t) - R \cdot \frac{p_3(t)}{L} - \mu \cdot \frac{p_6(t)}{J} \tag{5.34}$$

Now at second 1-junction we get

$$e_6(t) = e_5(t) - e_7(t) \tag{5.35}$$

This implies

$$\dot{p}_6(t) = \mu \cdot f_4(t) - \alpha \cdot e_8(t) \tag{5.36}$$

Where

$$e_8(t) = e_9(t) = k \cdot q_9(t) \tag{5.37}$$

$$f_4(t) = f_3(t) = \frac{p_3(t)}{L} \tag{5.38}$$

So

$$\dot{p}_6(t) = \mu \cdot \frac{p_3(t)}{L} - \alpha \cdot k \cdot q_9(t) \tag{5.39}$$

Now at a 0-Junction of the system we get

$$f_9(t) = f_8(t) - f_{10}(t) \tag{5.40}$$

$$\dot{q}_9(t) = \alpha \cdot f_7(t) - \frac{f_{11}(t)}{r} \tag{5.41}$$

Where

$$f_7(t) = f_6(t) = \frac{p_6(t)}{J} \tag{5.42}$$

$$f_{11}(t) = f_{12}(t) = \frac{p_{12}(t)}{m} \tag{5.43}$$

So

$$\dot{q}_9(t) = \alpha \cdot \frac{p_6(t)}{J} - \frac{1}{r} \cdot \frac{p_{12}(t)}{m} \tag{5.44}$$

At last for third 1-Junction, we write equations as

$$e_{12}(t) = e_{11}(t) + e_{13}(t) \tag{5.45}$$

With state variables and a constant acting force the Eq. (5.45) becomes

$$\dot{p}_{12}(t) = \frac{e_{10}(t)}{r} + W \tag{5.46}$$

$$\dot{p}_{12}(t) = \frac{k}{r} \cdot q_9(t) + W \tag{5.47}$$

The state space formulation is complete by using four equations as

$$\begin{bmatrix} \dot{p}_3(t) \\ \dot{p}_6(t) \\ \dot{q}_9(t) \\ \dot{p}_{12}(t) \end{bmatrix} = \begin{bmatrix} -R/L & -\mu/J & 0 & 0 \\ \mu/J & 0 & -\alpha \cdot k & 0 \\ 0 & 0 & \alpha/J & -1/r \cdot m \\ 0 & 0 & k/r & 0 \end{bmatrix} \cdot \begin{bmatrix} p_3(t) \\ p_6(t) \\ q_9(t) \\ p_{12}(t) \end{bmatrix} + \begin{bmatrix} 1 & 0 \\ 0 & 0 \\ 0 & 0 \\ 0 & 1 \end{bmatrix} \cdot \begin{bmatrix} F(t) \\ W \end{bmatrix} \tag{5.48}$$

The output at bond 12 can be measured easily if it is flow in this bond or effort at this junction. The flow in bond 12 relates to velocity of a mass. But in order to measure position, further integration of flow is required. In this bond p_{12} is generalized momentum and dividing by mass m will yield a velocity and integrating this velocity will result in displacement of mass. This integration increases the order of the system as one more equations need to be solved. Now let us define a new state variable q_{12} i.e., generalized displacement of bond as

$$\dot{q}_{12}(t) = \frac{p_{12}(t)}{m} \tag{5.49}$$

So state space formulation given in Eq. (5.48) is now represented with new state variable as

$$\begin{bmatrix} \dot{p}_3(t) \\ \dot{p}_6(t) \\ \dot{q}_9(t) \\ \dot{p}_{12}(t) \\ \dot{q}_{12}(t) \end{bmatrix} = \begin{bmatrix} -R/L & -\mu/J & 0 & 0 & 0 \\ \mu/J & 0 & -\alpha \cdot k & 0 & 0 \\ 0 & 0 & \alpha/J & -1/r \cdot m & 0 \\ 0 & 0 & k/r & 0 & 0 \\ 0 & 0 & 0 & 1/m & 0 \end{bmatrix} \cdot \begin{bmatrix} p_3(t) \\ p_6(t) \\ q_9(t) \\ p_{12}(t) \\ q_{12}(t) \end{bmatrix} + \begin{bmatrix} 1 & 0 \\ 0 & 0 \\ 0 & 0 \\ 0 & 1 \\ 0 & 0 \end{bmatrix} \cdot \begin{bmatrix} F(t) \\ W \end{bmatrix} \tag{5.50}$$

The output matrix for measuring both position and velocity is given as

$$\vec{y}(t) = \begin{bmatrix} 0 & 0 & 0 & 1 & 0 \\ 0 & 0 & 0 & 0 & 1 \end{bmatrix} \cdot \begin{bmatrix} p_3(t) \\ p_6(t) \\ q_9(t) \\ p_{12}(t) \\ q_{12}(t) \end{bmatrix} \tag{5.51}$$

If more variables need integration then order may be increase relatively. It depends upon sensors or required output to obtain a state space formulation. The state space in Eq. (5.48) fulfills the modeling requirement but for a specific output, the system needs more integration and hence the increases in system order. This can be achieved separately in simulations by obtaining velocity from this state space and further integrating it for position. A zero column in Eq. (5.50) also indicates the redundancy in the system caused by the sensor, which in this case is not adding new dynamics to the system but causing increase of order for specified variable output.

The major step in analytical formulation is to draw a bond graph representation appropriately with logical flow of power and causality assignments. When the bond graph has been drawn through steps given in the procedure then it will be easy to write equations. Equations at junctions are algebraic manipulations which at the end yield a state space representation of a model. This state space representation can be subjected to further analysis and design. In Example 5.7, we obtain two state space representations with 4th- and 5th-order model. The advantage of the 5th-order model is that we can obtain required output directly from model equations, but when linear analysis or controller design are performed, 4th-order model will prove itself a better and simple representation of the system. It is also an advantage that higher order differential equations are automatically represented in the system of equations through efforts and flows at respective junctions. However, these equations can be combined to obtain a single higher order system if needed.

5.5 20-Sim Software Tips

20-Sim software can directly generate system equations by drawing bond graph properly. The software then checks the necessary conditions of the model. If there are conflicts of causality, loops, or improper combination of elements, the software generates errors accordingly. The highlighted errors must be rectified and when there are no errors left, the modeling equations can be generated. The 20-Sim software can generate modeling equations in the presence of some warnings, but it is suggested that warnings must be taken care of for equations and properly understood in order to obtain a proper model of the system. By utilizing the frequency domain analysis tools, 20-Sim can also generate symbolic representation in state space formulation as well. Linearization in 20-Sim is only possible for single input and single output system.

Example 5.8 Equations through 20-Sim

Verify equations for a system given in Example 5.6 using 20-Sim Software

Solutions The state space formulation of the system is given in Eq. (5.25) as

$$\begin{bmatrix} \dot{p}_2(t) \\ \dot{q}_4(t) \end{bmatrix} = \begin{bmatrix} -\frac{B}{m} & -k \\ \frac{1}{m} & 0 \end{bmatrix} \cdot \begin{bmatrix} p_2(t) \\ q_4(t) \end{bmatrix} + \begin{bmatrix} 1 \\ 0 \end{bmatrix} \cdot [F(t)]$$
$$\vec{y}(t) = e_4(t) = \begin{bmatrix} 0 & k \end{bmatrix} \cdot \begin{bmatrix} p_2(t) \\ q_4(t) \end{bmatrix} \tag{5.52}$$

20-Sim generates the following equations as

Static equations:

```
F\p.e=F\effort;
```

Dynamic equations:

```
C\p.e=C\state/C\c;
C\p.f=m\state/m\i;
B\p.e=B\r * C\p.f;
m\p.e=F\p.e-(C\p.e+B\p.e);
```

System equations:

```
C\state=int (C\p.f, C\state_initial);
m\state=int (m\p.e, m\state_initial);
```

Static and state equations represent input and state vectors respectively. Dynamic equations represent the state equations but their number may be equal or more than the order. By representing the system equations in terms of derivatives on the left hand side rather than integrals on the right hand side, we get the state space representation. Backslash with variables show parametric assignment and forward slash is used for division. C\state and m\state represent the $q_4(t)$ and $p_2(t)$ respectively. Looking at the first system equation and substituting C\p.f accordingly, we obtain the 2nd equation in state space formulation as in Eq. (5.24). Now substituting m\p.e in the 2nd system equation we get the 1st equation in state space representation i.e., Eq. (5.23). These equations are the same as the derived equation in Example 5.6. Using "Model Linearization" from frequency domain tools, we can get state space representation directly in 20-Sim environment as well as transfer function of the system (with default numerical values of parameters). In addition there are 14 removed equations after algebraic manipulation to obtain dynamic equations, which are not needed in the process.

Problems

P5.1 Draw a bond graph of each system and represent its equation in state space formulation and verify it using 20-Sim software.

P5.2 Use state space formulation separately for both cases of Example 5.2 using 20-Sim and verify equations of 20-Sim with equations manually calculated of both representations.

P5.3 A good practice for drawing bond graph and generating an equation is to study the related examples and material in the literature. Ref. [1], "System Dynamics—Modeling, Simulation, and Control of Mechatronic Systems" by *D C Karnopp, D L Margolis & R C Rosenberg* is a wonderful resource for bond graph methodology in detail. Draw bond graph for each example and exercise of Chap. 4, of this text and obtain equations manually and verify it using 20-Sim software.

P5.4 All examples in this chapter and P5.1 have been implemented in 20-Sim software with both bond graph methodology and block diagram. It is possible for many examples to directly obtain equations from block diagrams rather than converting these into bond graphs. Implement these examples/P5.1 in 20-Sim using block diagrams and obtain equations and compare with state space representation in this form.

P5.5 Obtain transfer functions of the systems in the examples in Chap. 5 and P5.1 and compare with the transfer function obtained from 20-Sim linearization tool. Note that the numerator of the transfer function depends upon the output chosen for the purpose whereas the denominator remains the same regardless of the output variable.

References

1. Karnopp, Dean C., Donald L. Margolis, and Ronald C. Rosenberg. 2012. *System Dynamics—Modeling and Simulation of Mechatronics Systems*, 5th ed. New York: Wiley.
2. 20-Sim Software, www.20sim.com
3. Asif M. Mughal, and Kamran Iqbal. 2013. Bond Graph Modeling and Optimal Control Design for Physiological Motor Control System. *International Journal of Modeling and Simulation*, 33(2):93–101.

Chapter 6
Advance Bond Graph Modeling

We obtain the biggest advantage of bond graph when a system is composed of different energy domains. In this chapter, we will cover some examples of bond graph modeling in different energy systems with advanced and more practical concepts in addition to the details covered in the last chapter. In the last chapter, basic bond graph formulation and state space representation were discussed. In this chapter, we will discuss the concepts of algebraic loops, additional two port elements known as fields, the effect of derivative causality, and bond graph in the vector notation.

6.1 Algebraic Loops

Modeling of different systems has different types of problems, sometimes a simple model can lead to a difficult mathematical problem. It often occurs that due to our assumptions in modeling, a simple problem leads to a difficult one and on the other hand a difficult problem yields to be a simpler solution. Figure 6.1 shows a simple circuit where R_2 is a current limiting resistor to the load. Due to this resistor, a loop is created in the circuitry that makes the state space formulation of the system more tedious and dependent even though there is no differential causality in the system. Figure 6.2 shows the bond graph of the system.

The state equations for second order system with two states q_3 and p_7 are

$$\dot{q}_3 = f_4 = f_5 + f_7 \tag{6.1}$$

The flow (current) in the first 1-junction divides into limiting resistor and load resistor where more simplification leads to

A.M. Mughal, *Real Time Modeling, Simulation and Control of Dynamical Systems*,
DOI 10.1007/978-3-319-33906-1_6

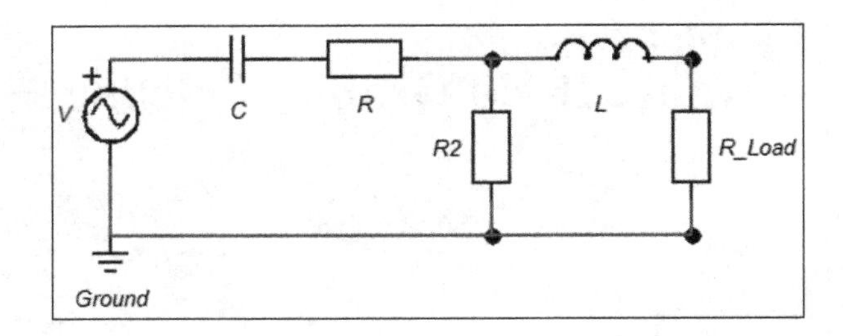

Fig. 6.1 Electrical circuit with an algebraic loop

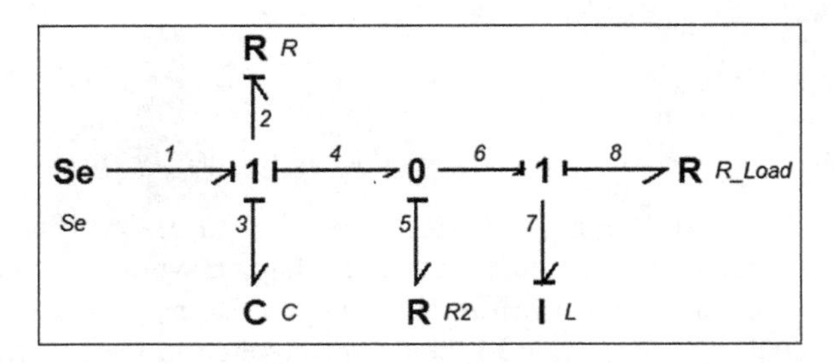

Fig. 6.2 Bond graph of a system given in Fig. 6.1

$$\dot{q}_3 = \frac{e_5}{R_2} + \frac{1}{L} \cdot p_7 \tag{6.2}$$

The effort (voltage) across resistor R_2 distributes as

$$e_5 = e_4 = e_1 - e_2 - e_3 \tag{6.3}$$

$$\dot{q}_3 = \frac{1}{R_2}\left(V - R \cdot f_2 - \frac{1}{C} \cdot q_3\right) + \frac{1}{L} \cdot p_7 \tag{6.4}$$

Now flow in bond 2 through resistor R is the same as the flow in bond 3, which is being solved. This creates an algebraic loop.

$$\dot{q}_3 = \frac{1}{R_2}\left(V - R \cdot \dot{q}_3 - \frac{1}{C} \cdot q_3\right) + \frac{1}{L} \cdot p_7 \tag{6.5}$$

Now moving the term $\frac{R}{R_2} q_3$ on the left hand side solves this loop as

$$\dot{q}_3\left(1 + \frac{R}{R_2}\right) = \frac{1}{R_2}\left(V - \frac{1}{C} \cdot q_3\right) + \frac{1}{L} \cdot p_7 \tag{6.6}$$

Now solving for a second variable, we get effort across inductor as

$$\dot{p}_7 = e_6 - e_8 = e_4 - f_8 \cdot R_L \tag{6.7}$$

We get

$$\dot{p}_7 = \left(V - R \cdot \dot{q}_3 - \frac{1}{C} \cdot q_3\right) - \frac{R_L}{L} \cdot p_7 \tag{6.8}$$

By substituting a complete state equation from Eq. (6.6), we yield

$$\dot{p}_7 = \left(V - \frac{R \cdot \left(\frac{1}{R_2}(V - \frac{1}{C} \cdot q_3) + \frac{1}{L} \cdot p_7\right)}{\left(1 + \frac{R}{R_2}\right)} - \frac{1}{C} \cdot q_3\right) - \frac{R_L}{L} \cdot p_7 \tag{6.9}$$

The cumbersome state equation now requires careful algebra in order to get the state space representation as

$$\dot{q}_3 = -\frac{1}{(R + R_2)C} \cdot q_3 + \frac{R_2}{(R + R_2)L} \cdot p_7 + \frac{1}{(R + R_2)} \cdot V \tag{6.10}$$

$$\dot{p}_7 = -\frac{1}{\left(1 + \frac{R}{R_2}\right)C} \cdot q_3 - \frac{R + \left(1 + \frac{R}{R_2}\right) \cdot R_L}{\left(1 + \frac{R}{R_2}\right)L} \cdot p_7 + \frac{1}{\left(1 + \frac{R}{R_2}\right)} \cdot V \tag{6.11}$$

More than one algebraic loop may result in more complex algebraic operations in order to obtain state space equations. In this model if current limiting from the source is taken care at just after source and before capacitor, the model will be simplified with a zero junction and a one junction without algebraic loop. This can further be simplified as a source of flow with a one junction as shown in Fig. 6.3 but it will introduce a differential causality in the system. However, it must be understood that this arrangement may not be possible with all the modeling assumptions e.g., if there is a transformer in the system or the voltage drop across capacitors needs attention, etc.

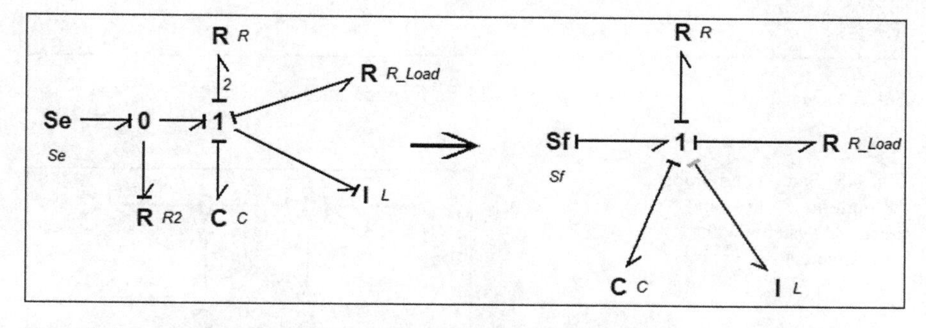

Fig. 6.3 Bond graph simplification of system in Fig. 6.1 but causing derivative causality

6.2 Derivative Causality

A system with derivative causality is actually a redundant system in terms of the number of equations. The state of energy-storing element is dependent upon other states which creates linearly dependent rows in the matrix. For example, electrical capacitance in parallel or mechanical spring in series leads to derivative causalities as shown in Fig. 6.4. Both systems have same bond graph representation as shown in Fig. 6.5.

We write state space equation of the bond graph with two state variables q_4 and q_5. The effort at 1-junction expresses as

$$e_1(t) = e_2(t) + e_3(t) \tag{6.12}$$

Which further solve as

$$S_e = f_2 \cdot R + \frac{1}{C_1} \cdot q_4 \tag{6.13}$$

The effort at 0-junction can be expressed as either q_4 not at q_5 due to differential causality at port 5. We now express the state equations by equating flow in 1-junction i.e., $f_2 = f_3$ and

$$f_3 = f_4 + f_5 = \dot{q}_4 + \dot{q}_5 \tag{6.14}$$

Substituting Eq. (6.13) in Eq. (6.14) yields

$$\dot{q}_4 = \frac{1}{R} S_e - \frac{1}{R \cdot C_1} \cdot q_4 - \dot{q}_5 \tag{6.15}$$

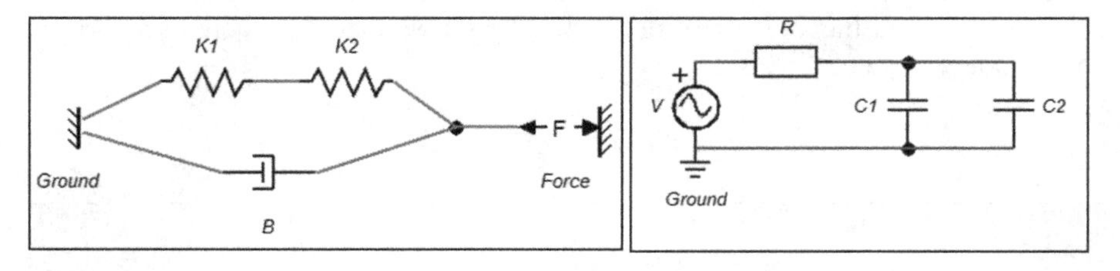

Fig. 6.4 Mechanical and electrical systems with derivative causality

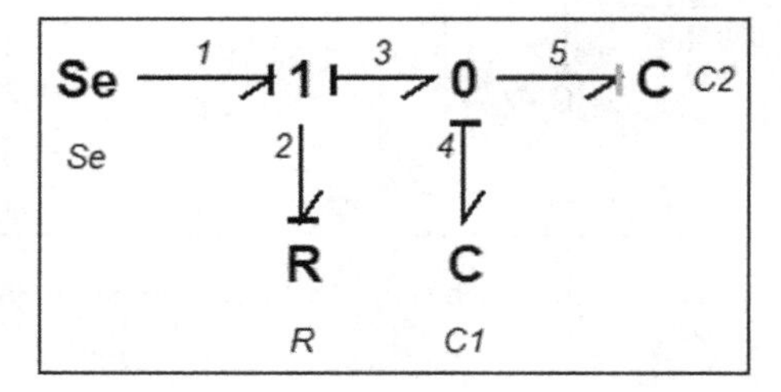

Fig. 6.5 Same bond graph for both mechanical and electrical systems with derivative causality

Now effort at 0-junction is given as

$$e_4(t) = e_5(t) \tag{6.16}$$

$$\frac{q_4(t)}{C_1} = \frac{q_5(t)}{C_2} \tag{6.17}$$

Now differentiating Eq. (6.17) with time implies to

$$\dot{q}_5 = \frac{C_2}{C_1} \cdot \dot{q}_4 \tag{6.18}$$

Substituting Eq. (6.18) in Eq. (6.15) gives us first state equation as

$$\dot{q}_4 = \frac{1}{1 + {}^{C_2}/_{C_1}} \left(\frac{1}{R} S_e - \frac{1}{R \cdot C_1} \cdot q_4 \right) \tag{6.19}$$

The second state equation is linearly dependent upon first state equation as given in Eq. (6.18)

$$\dot{q}_5 = \frac{{}^{C_2}/_{C_1}}{1 + {}^{C_2}/_{C_1}} \left(\frac{1}{R} S_e - \frac{1}{R \cdot C_1} \cdot q_4 \right) \tag{6.20}$$

Now the state space representation is given as

$$\begin{bmatrix} \dot{q}_4 \\ \dot{q}_5 \end{bmatrix} = \frac{1}{1 + {}^{C_2}/_{C_1}} \cdot \left(-\frac{1}{R \cdot C_1} \cdot \begin{bmatrix} 1 & 0 \\ {}^{C_2}/_{C_1} & 0 \end{bmatrix} \cdot \begin{bmatrix} q_4 \\ q_5 \end{bmatrix} + \frac{1}{R} \cdot \begin{bmatrix} 1 \\ {}^{C_2}/_{C_1} \end{bmatrix} \cdot S_e \right) \tag{6.21}$$

The A matrix has one column zero and that is due to the linear dependence, which is the result of derivative causality. This simply means that one state equation is enough to solve the system and the other equation is simply an algebraic multiple of another equation. The derivative causality introduces an algebraic loop in the systems which are required to be solved by manipulating variable in the formulation. Let us consider a circuit in Fig. 6.6, where current flow in C_2 is dependent upon C_1 and both are related to different state variables.

The bond graph of the circuit given in Fig. 6.6 is given in Fig. 6.7 with causality assignment at C_1, however, causality assignment in this case may be changed either to C_1 or C_2. In this case the independent state variable is q_2 and the dependent state variable is q_4.

In this algebraic loop, we can only get one state equation for q_2 and the other state q_4 can be obtained at output. Problem P6.5 elaborates more for the complete state space representation which requires differentiation of source.

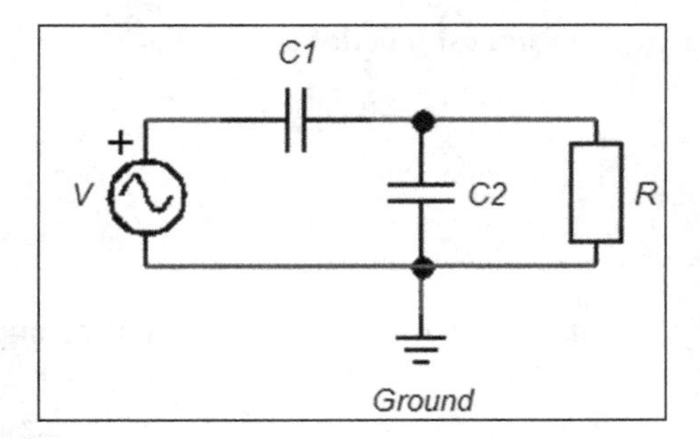

Fig. 6.6 Two capacitor in derivative causality configurations

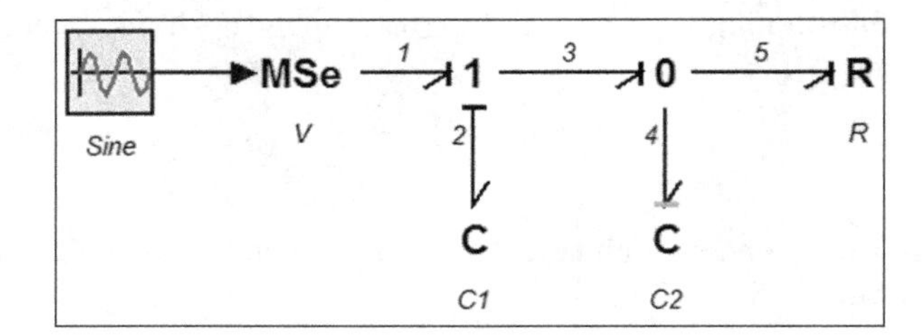

Fig. 6.7 Bond graph of an electrical circuit given in Fig. 6.6 given Fig. 6.7

$$\dot{q}_2 = \frac{1}{R} \cdot \left(V - \frac{q_2}{C_1} \right) \cdot \frac{1 - {}^{C_2}\!/_{C_1}}{1 + {}^{C_2}\!/_{C_1}} \tag{6.22}$$

$$q_4 = C_2 \cdot V + q_2 \cdot \frac{C_2}{C_1} \tag{6.23}$$

6.3 Fields

Basic 1-port elements may need higher representation in modeling, e.g., a spring is attached with both ends exerting force or capacitors banks in electrical circuits, different configurations of inductor as well as modulated transformers and gyrators with own feedbacks. These basic elements can be represented as Fields, in which a 1-port element is connected with two or more ports with the same physical law. The constitutive algebraic law will be represented in matrix form for generalized variables. Let us consider two similar examples in mechanical and electrical systems with two generalized capacitances with two sources of efforts as shown in Fig. 6.8 and Fig. 6.9:

The respective bond graph for both systems is same and shown in Fig. 6.10

Both generalized capacitors have differential causality and it makes it more difficult to obtain the equations. The effort, flow, or displacement functions in C_1 are dependent upon C_2 and vice versa. This situation can easily be represented with Field capacitors with two ports as shown in Fig. 6.11.

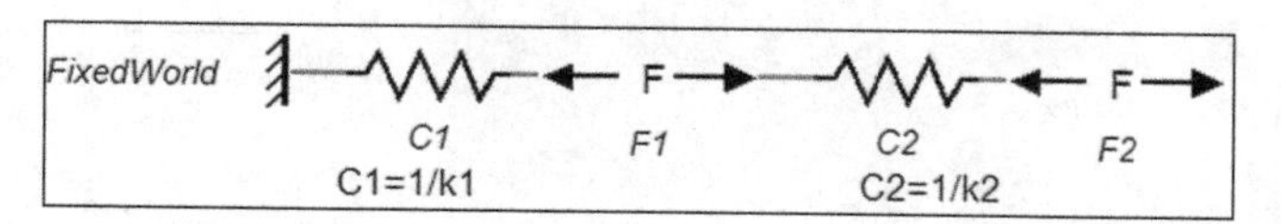

Fig. 6.8 Two springs with two forces

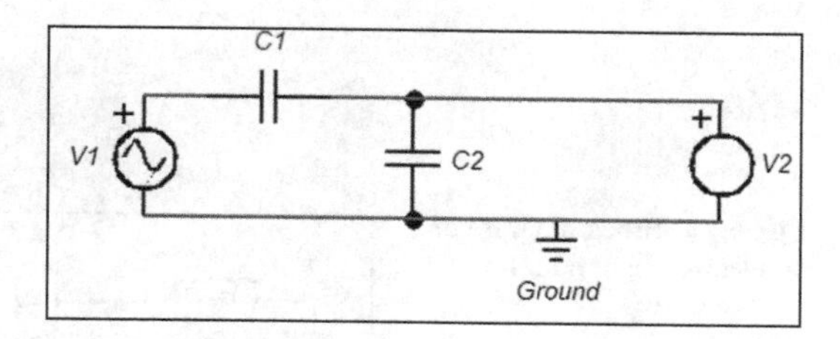

Fig. 6.9 Two capacitors with voltage sources

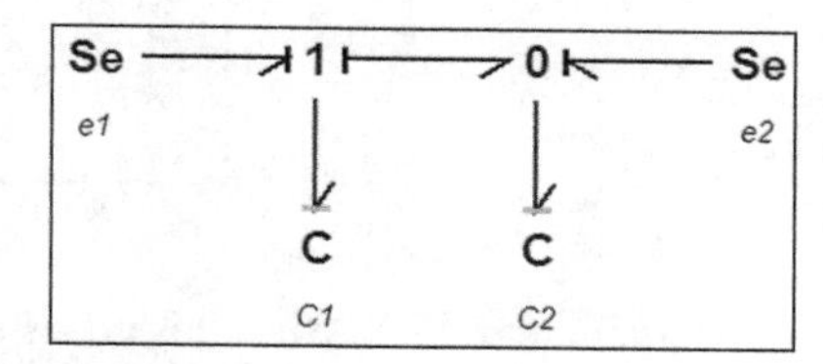

Fig. 6.10 Bond graph of two compliances with two effort sources given in Figs. 6.8 and 6.9

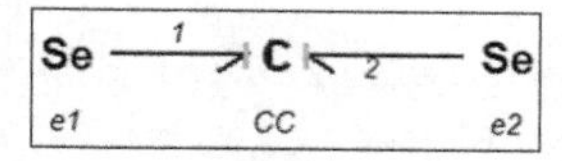

Fig. 6.11 Bond graph of 2-port field compliances

The constitutive equation will be given as:

$$\begin{bmatrix} e_1 \\ e_2 \end{bmatrix} = \begin{bmatrix} \frac{1}{C_1} + \frac{1}{C_2} & \frac{1}{C_2} \\ 0 & \frac{1}{C_2} \end{bmatrix} \cdot \begin{bmatrix} q_1 \\ q_2 \end{bmatrix} \tag{6.24}$$

Similarly, an induction transformer action represents better as I-Field element with a mutual inductance value given as:

The Fig. 6.12 above shown here is with a mixed causality assignment, which is also appropriate according to the requirement of the system. The constitutive equations of this system are given as

$$\begin{bmatrix} \lambda_1 \\ \lambda_2 \end{bmatrix} = \begin{bmatrix} L_1 & M_{12} \\ M_{12} & L_2 \end{bmatrix} \cdot \begin{bmatrix} i_1 \\ i_2 \end{bmatrix} \tag{6.25}$$

The word *C* with an *I* represents that this is *I*-field element, whereas *C* is a denotation of field in 20-Sim software. This can also be represented as bold **I** or

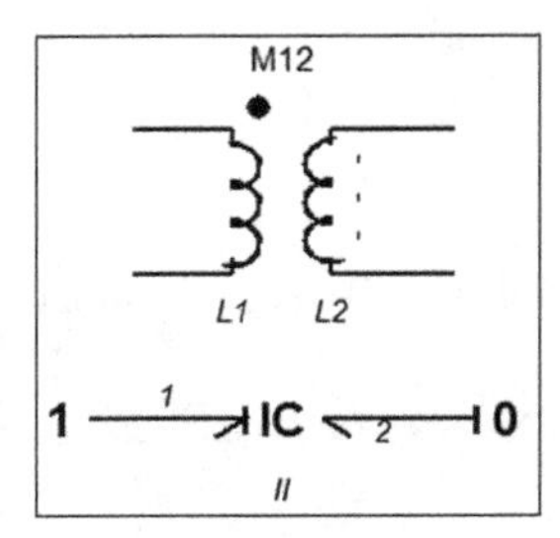

Fig. 6.12 Induction transformer and its bond graph with 2-port field inductor

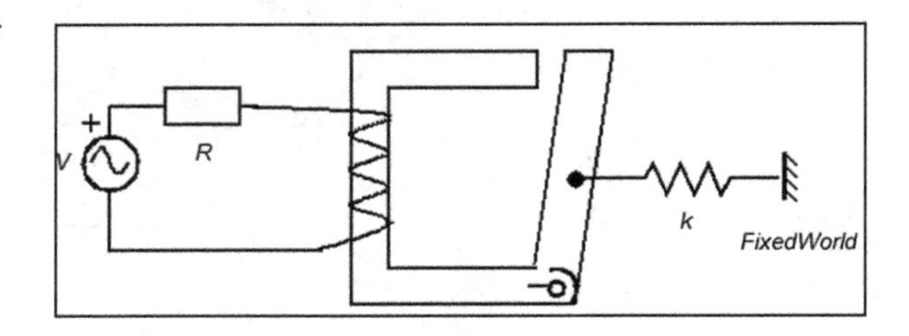

Fig. 6.13 Block diagram of an electromagnetic relay

2-*I* element, etc. In 20-Sim, it is necessary to change in code to represent the II-field or resistive fields given as

```
state1 = int (p1.e);
  state2 = int (p2.e);
  [ p1.f; p2.f ] = ic * [ state1; state2 ];
```

The main advantage of this port-scheme is significant when there are different energy systems with transformer or gyrator actions through fields, e.g., an electromagnetic relay which consists of three primary systems: electrical, magnetic, and mechanical (Fig. 6.13).

A normally open electromagnetic relay converts electrical energy into magnetic energy, which closes the circuits till the voltage is being supplied. As soon as voltage across the windings is released, a mechanical spring pushes the mass and thus opens the relay. The magnetic force equates the rotation, damping action, and exerting force in the mechanical spring. A gyrator is needed to represent conversion of electrical energy into magnetic energy, and a gyrator modulus is the number of turns in the coil. In order to avoid detailed magnetic circuits, we can represent the magnetic resistance on a zero junction followed by a *C*-field capacitor connecting the spring and mechanical assembly together. A simplified bond graph of the system is represented as (Fig. 6.14):

Remember this is a simplified version of bond graph for an electromagnetic relay. A detailed discussion and better model for this electromagnetic relay is given in ref [1]. This system has three state variables, two displacement variables q_6, and q_7 and a momentum p_8.

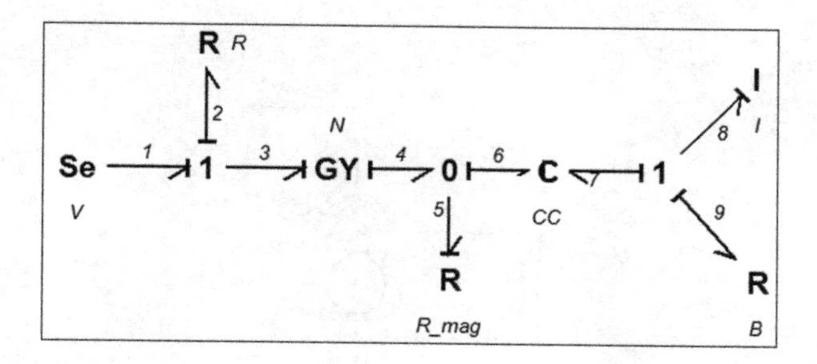

Fig. 6.14 Bond graph of an electromagnetic relay given in Fig. 6.13

$$\begin{bmatrix} \dot{q}_6(t) \\ \dot{q}_7(t) \\ \dot{p}_8(t) \end{bmatrix} = \begin{bmatrix} -\frac{1}{C_6}\left(\frac{R}{N^2}+\frac{1}{R_{\text{mag}}}\right) & -\frac{1}{C_{67}}\left(\frac{R}{N^2}+\frac{1}{R_{\text{mag}}}\right) & 0 \\ 0 & 0 & \frac{1}{I} \\ -\frac{1}{C_{76}} & -\frac{1}{C_7} & -\frac{B}{I} \end{bmatrix} \cdot \begin{bmatrix} q_6(t) \\ q_7(t) \\ p_8(t) \end{bmatrix} + \begin{bmatrix} \frac{1}{N} \\ 0 \\ 0 \end{bmatrix} \cdot [V(t)] \tag{6.26}$$

In this equation C_{67} is a mutual compliance of magnetic circuit with mechanical assembly and C_{76} is a compliance observed by mechanical system from magnetic behavior. Whereas, C_6 and C_7 are reluctance of electromagnet and reciprocal of spring stiffness respectively.

6.4 Series Motor

A block diagram of a series motor with armature and field voltages and current is shown in Fig. 6.15 with ideal bond graph.

The bond graph of a network with a series motor in connection with loads through a transformer is shown in Fig. 6.16. The input to the transformer is the rotational effort (torque) and angular rate of motor shaft which drives the load of mass m with damping in rope B_{out}. There are field R_f, armature R_a, and output resistances R of series motor respectively, inductance at output, spring and damper at shaft end are given by I, k, and B respectively. In this circuit an IC field represents a capacitor and inductor at respective bonds as well as parasitic capacitance and inductance in the windings. In IC field, effective parasitic capacitance and inductance along with individual elements are represented in a vector notation given as

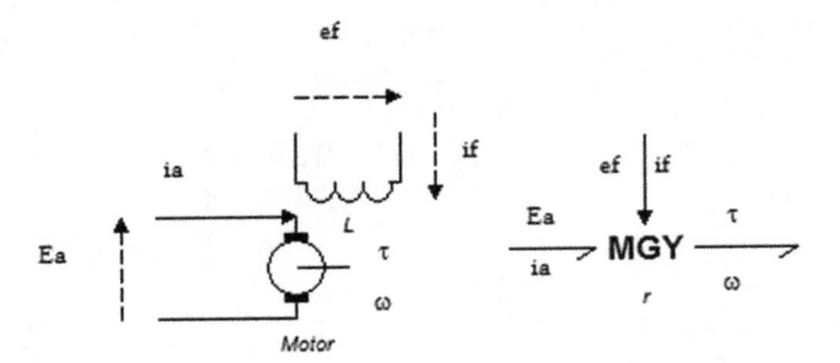

Fig. 6.15 A series motor with armature and field sources and bond graph

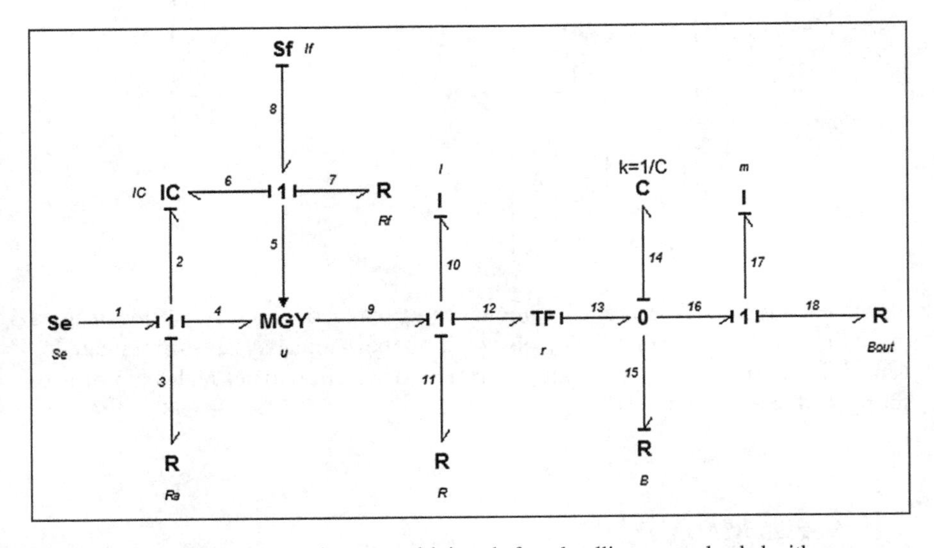

Fig. 6.16 A network with a series motor driving shaft and pulling a rope loaded with mass

$$\begin{bmatrix} f_6 \\ q_2 \end{bmatrix} = \begin{bmatrix} \dfrac{1}{I_6} & a_{62} \\ a_{26} & \dfrac{1}{C_2} \end{bmatrix} \cdot \begin{bmatrix} p_6 \\ q_2 \end{bmatrix} \tag{6.27}$$

If $a_{62} = a_{26} = 0$ then the effect of parasites, windage, or air gap are ignored, and it represents a simple inductor element at port 6 and a simple capacitive element at port 2. Otherwise, $a_{62} = \frac{1}{C_{62}}$ is a parasitic capacitance faced at field windings from armature and $a_{26} = \frac{1}{L_{26}}$ is a mutual inductance faced at armature due to field windings. Both energy storing variables p_6 and q_2 in IC field will also be state variables in the system equations.

6.5 Resistive Fields

The causality assignment in a resistance element can be described as resistance causality for measuring effort and conductance causality for measuring flow at the port (Table 6.1).

Resistive fields are classified into resistive form and conductance form due to causality assignments as shown in Table 6.2:

In many electrical circuits, it is often a requirement to measure the combination of voltages, or currents in some branches. So the output is the combination of voltages in different loops as current is the combination of currents in different nodes. Resistance element provides us with flexibility for measurement without any burden of integral and differential causality. A pure resistive circuit output can be formed in a matrix form for measurement of effort or flow just like network diagram. In addition, in many electrical and mechanical systems, mixed causality of resistive elements is a more practical scenario. In mixed causality assignments, both resistance and conductance forms are applicable at a time and providing measurement of both effort and flow just like a hybrid network diagram. The examples of matrices for different forms are given in Table 6.3.

Table 6.1 Resistance and conductance causality

Resistance causality	⊢⇀ R	$e(t) = R \cdot f(t)$
Conductance causality	⇀⊣ R	$f(t) = \frac{e(t)}{R}$

Table 6.2 Resistance and conductance form with constitutive laws

Resistance form	1 2 3 4 R n	$e(t) = \varphi_R\{f_1(t), f_2(t), f_3(t), \ldots f_n(t)\}$
Conductance form	1 2 3 4 R n	$f(t) = \varphi_R^{-1}\{e_1(t), e_2(t), e_3(t), \ldots e_n(t)\}$

Table 6.3 Matrix notation for measuring of effort or flow in resistive fields

Resistance form	$\begin{bmatrix} e_1 \\ e_2 \\ e_3 \end{bmatrix} = \begin{bmatrix} R_1 & R_{12} & 0 \\ 0 & R_2 & R_{23} \\ R_{31} & R_{32} & R_3 \end{bmatrix} \cdot \begin{bmatrix} f_1 \\ f_2 \\ f_3 \end{bmatrix}$
Conductance form	$\begin{bmatrix} f_1 \\ f_2 \\ f_3 \end{bmatrix} = \begin{bmatrix} 1/R_1 & 0 & 0 \\ 0 & 1/R_2 & 1/R_2 + 1/R_3 \\ 1/R_{eq} & 0 & 1/R_3 \end{bmatrix} \cdot \begin{bmatrix} e_1 \\ e_2 \\ e_3 \end{bmatrix}$
Mixed form	$\begin{bmatrix} f_1 \\ e_2 \\ e_3 \end{bmatrix} = \begin{bmatrix} 1/R_1 & 0 & 1/R_2 + 1/R_3 \\ 1 & R_2 & 0 \\ 0 & 0 & R_3 \end{bmatrix} \cdot \begin{bmatrix} e_1 \\ f_2 \\ f_3 \end{bmatrix}$

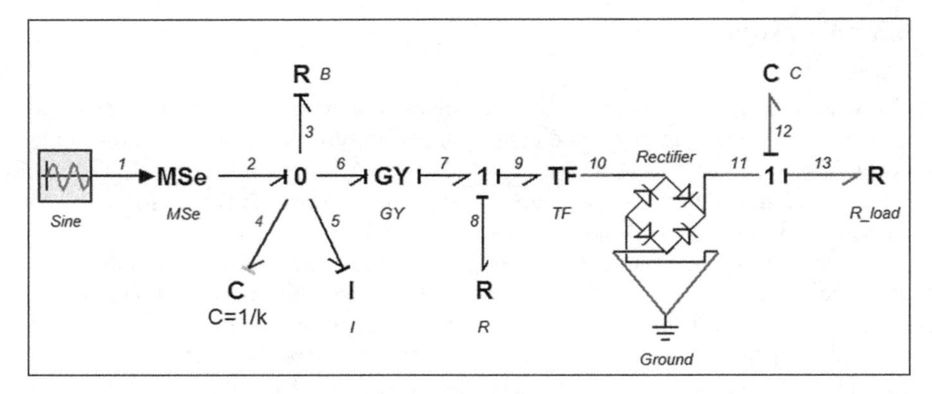

Fig. 6.17 AC generator network with voltage rectification

6.6 Bridge Circuit

An electrical generator produces torque gyrated into electricity in AC form. The AC electrical energy is stepped down using a transformer followed by a bridge rectifier to convert into DC. The DC electrical energy drives the load R_{Load} with a capacitance C in series to take care of ripples. The bond graph shown in Fig. 6.17 is a conventional bond graph with a rectifier element. The rectifier is converting AC into DC peaks with a $e_{11} = |e_{10}|$

6.7 Combination of Elements

In Chap. 4, problem 4, we wrote a constitutive law between combinations of gyrators and transformers with all possible choices of causality. However, in particle all possible choices of causality may not be possible, e.g., a 0-junction followed by a transformer, followed by another transformer and again followed by 0-junction is possible and simplified as a single transformer between 0-junctions. This choice is not possible for two gyrators with a 0-junction at each end. For a mixed combination of gyrators and transformers with either of 0-junction or 1-junction at any end, all possible choices of causalities are also not possible. Figure 6.18 shows the only possible cases of causalities with 0-junctions at both ends.

In modeling with bond graph, it is also possible to get a model where transformers and gyrators have either 0-junction or 1-junction between them with no other element connected at this junction. Obviously this junction will be reduced with respective constitutive laws. Sources connected with gyrators also reduce the bonds by forming new constitutive laws given in Table 6.4:

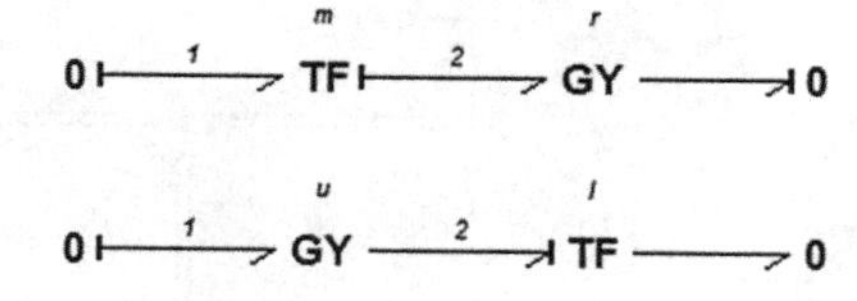

Fig. 6.18 Possibilities of transformer and gyrator in series

Table 6.4 Sources followed by a gyrator element

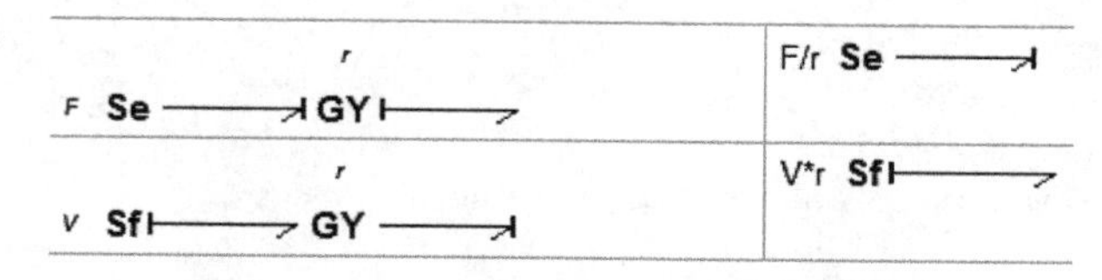

r F Se GY	F/r Se
r V Sf GY	V*r Sf

In some literature, other single port elements with gyrators or transformers are replaced for simplicity, e.g., a gyrator followed by an inductor can replace both gyrator and inductor as equivalent capacitor. Also, a capacitor followed by a gyrator is simplified as an inductor. A transformer followed by a capacitor or inductor will respectively replace with capacitor or inductor. But these simplifications often lead to confusion relating to second ports of the two-port elements and needs more attention while formulating constitutive laws as well as state space equation. It is therefore not necessary to follow these simplifications. A resistor followed by a transformer or gyrator remains a resistor with appropriate constitutive law (P6.10).

6.8 Vector Bond Graphs

Electrical networks or mechanical networks for fluid flow and other applications have many similar ports representing the same constitutive laws between elements as shown in Fig. 6.19. There may be interactions between elements that can also be represented such as mutual inductance, two-way spring action just as C-fields and I-fields. The constitutive laws remain the same but appear in matrix notations.

The figure shows a vector bond graph. Let us consider it is a 3×3 system for each single port element. This implies, C, I, and R are 3×3 matrices with diagonal elements representing values of each individual elements and non-diagonal values represent interaction elements given as

$$C = K = \begin{bmatrix} k_{11} & k_{12} & k_{13} \\ k_{21} & k_{22} & k_{23} \\ k_{31} & k_{32} & k_{33} \end{bmatrix} \quad I = M = \begin{bmatrix} m_{11} & m_{12} & m_{13} \\ m_{21} & m_{22} & m_{23} \\ m_{31} & m_{32} & m_{33} \end{bmatrix} R = \begin{bmatrix} R_{11} & R_{12} & R_{13} \\ R_{21} & R_{22} & R_{23} \\ R_{31} & R_{32} & R_{33} \end{bmatrix}$$

If non-diagonal elements are zero, then diagonal elements only represent the self values. In this system, it should be noted that the values of any element $[a_{ij}]$ represent the interaction of *ith* element with *jth* element in a matrix. For example, k_{12} is the combined compliance of two capacitances C_1 and C_2 just like field

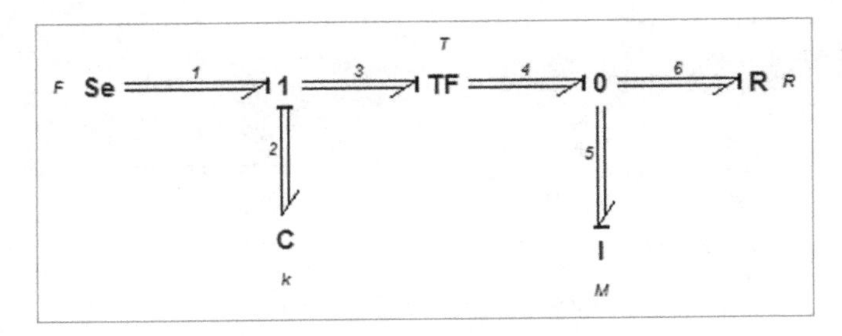

Fig. 6.19 A vector bond graph of a simple network

capacitance. The equations of the system have the same methods except vectors, and matrices will be used. In this example, the state vector consists of two vector elements, $\vec{q_2}$ and $\vec{p_5}$, where each element in state vector can be represented as

$$\vec{q_2} = \begin{bmatrix} q_{21} \\ q_{22} \\ q_{23} \end{bmatrix} \quad \vec{p_5} = \begin{bmatrix} p_{51} \\ p_{52} \\ p_{53} \end{bmatrix} \tag{6.28}$$

$$\vec{x}(t) = \begin{bmatrix} \vec{q_2}(t) \\ \vec{p_5}(t) \end{bmatrix}$$

The causality assignment of vector bond graph is the same; the methodology of obtaining equations at each junction also remains the same. The first set of equations is obtained as

$$\vec{e_3}(t) = \vec{e_1}(t) - \vec{e_2}(t) \tag{6.29}$$

$$\vec{e_3}(t) = T \cdot \vec{e_4}(t) \tag{6.30}$$

$$\vec{f_4}(t) = \vec{f_5}(t) + \vec{f_6}(t) \tag{6.31}$$

The solution to be found is $\vec{\dot{x}} = \begin{bmatrix} \vec{\dot{q}_2} \\ \vec{\dot{p}_5} \end{bmatrix}$, where $\vec{\dot{q}_2}$ and $\vec{\dot{p}_5}$ are two independent vectors. From the figure and set of equations, we can get following equations as

$$\vec{\dot{q}_2} = \vec{f_2}(t) = \vec{f_3}(t) \tag{6.32}$$

$$\vec{\dot{q}_2} = \vec{f_3}(t) = T^{-1} \cdot \vec{f_4}(t) \tag{6.33}$$

$$\dot{\vec{q}}_2 = T^{-1} \cdot \left(\vec{f_5}(t) + \vec{f_6}(t) \right) \tag{6.34}$$

Where

$$\vec{f_5}(t) = M^{-1} \cdot \vec{p_5} \tag{6.35}$$

and

$$\vec{f_6}(t) = R^{-1} \cdot \vec{e_6}(t) \tag{6.36}$$

$$\vec{e_6}(t) = \vec{e_4}(t) = T^{-1} \cdot \vec{e_3}(t) \tag{6.37}$$

$$\vec{e_3}(t) = \vec{F}(t) - K^{-1} \cdot \vec{q_2} \tag{6.38}$$

$$\vec{f_6}(t) = R^{-1} \cdot T^{-1} \left(\vec{F}(t) - K^{-1} \cdot \vec{q_2} \right) \tag{6.39}$$

Substituting Eqs. (6.35) and (6.39) in Eq. (6.34) gives us

$$\dot{\vec{q}}_2 = T^{-1} \cdot \left(M^{-1} \cdot \vec{p_5} + R^{-1} \cdot T^{-1} \left(\vec{F}(t) - K^{-1} \cdot \vec{q_2} \right) \right) \tag{6.40}$$

Similarly for second variable vector $\dot{\vec{p}}_5$ we have

$$\dot{\vec{p}}_5 = \vec{e_5}(t) = \vec{e_6}(t) \tag{6.41}$$

So we can easily get

$$\dot{\vec{p}}_5 = T^{-1} \left(\vec{F}(t) - K^{-1} \cdot \vec{q_2} \right) \tag{6.42}$$

The combined state space of the system is given as

$$\begin{bmatrix} \dot{\vec{q}}_2 \\ \dot{\vec{p}}_5 \end{bmatrix} = \begin{bmatrix} -(TRT)^{-1} \cdot K^{-1} & (MT)^{-1} \\ -T^{-1} \cdot K^{-1} & 0 \end{bmatrix} \cdot \begin{bmatrix} \vec{q_2}(t) \\ \vec{p_5}(t) \end{bmatrix} + \begin{bmatrix} (TRT)^{-1} \\ T^{-1} \end{bmatrix} \cdot \vec{F}(t) \tag{6.43}$$

The state vector and input $\vec{F}(t)$ are 6×1 and 3×1 vectors respectively for the overall sixth order system given in Eq. (6.43).

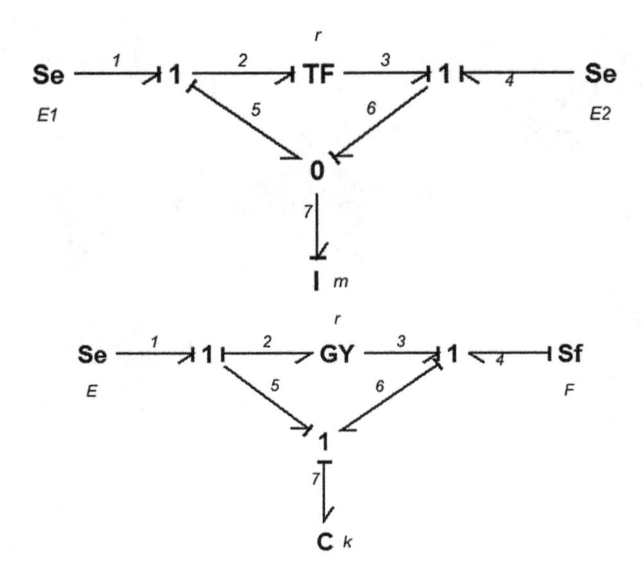

Fig. 6.20 Algebraic loops with two port elements (**a**) transformer network (**b**) Gyrator network

Problems

P6.1 Find the state space representation of loop circuits given in Fig. 6.20. Is it possible to obtain different state space by reversing the loop (i.e., counterclockwise)?

P6.2 Draw a block diagram of the bond graph of a system given in 6.16 with given details. Find state space representation to measure the position of mass attached to the rope and maximum force acting on the rope.

P6.3 Formulate a state space relation of the bridge circuit given in Fig. 6.17. Is there any difficulty in obtaining the output equation for measuring voltage across load resistor? Is it possible to obtain the transfer function of the system?

P6.4 Find the bond graph diagram of the following circuits in Fig. 6.21 and explain how many choices of causalities are possible. Write state space equations of the system of all possible choices. Are the state space representations different or equivalent? Explain

P6.5 In order to represent state space with the two variables of systems given in Fig. 6.6, an effort source can be represented as $V(t) = V_m \sin(\omega t)$. Now differentiation of this is possible, which may lead to state space representation with two variables. Formulate a state space representation and give a transfer function from each state as output to the input $V(t)$.

P6.6 Capacitor bank in parallel are used for amplification in voltage in certain applications. A simple circuit is shown in Fig. 6.22 ith n number of capacitors. Derive a nth order state space representation and discuss the structure of A and B matrices.

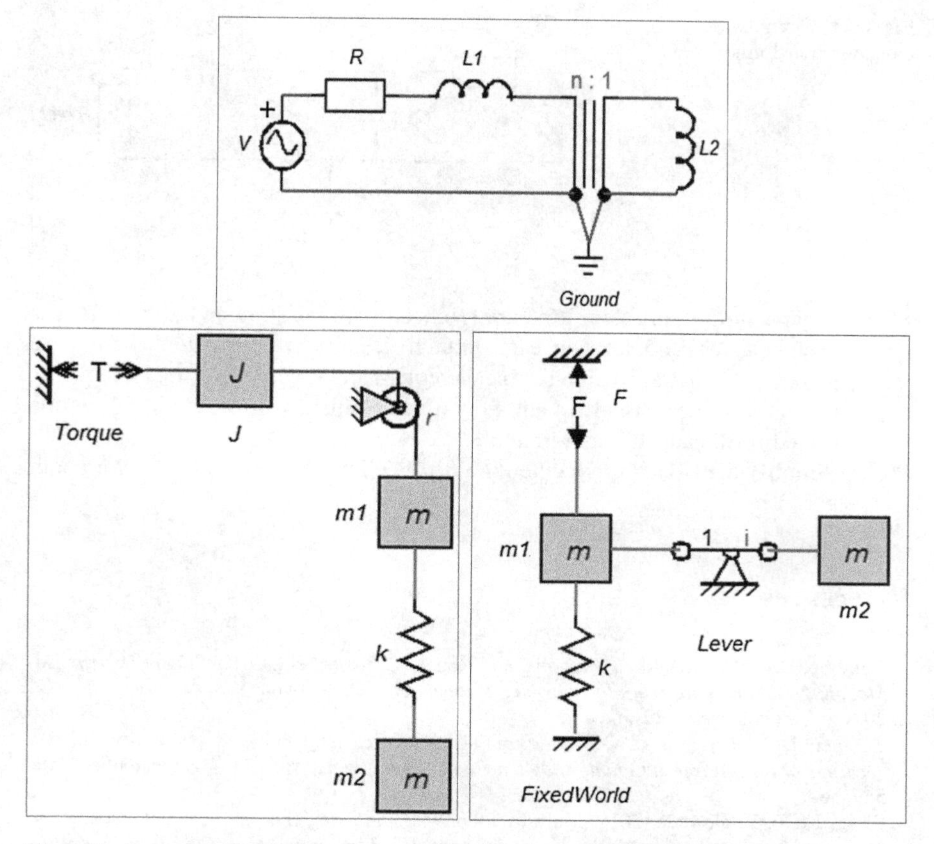

Fig. 6.21 Systems with different choices of causality assignments

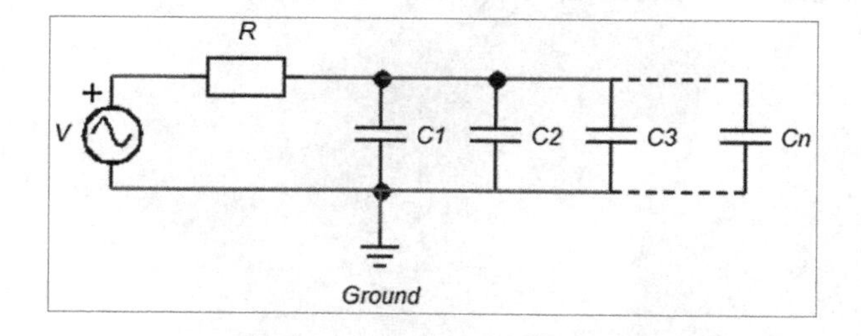

Fig. 6.22 Capacitor bank network

P6.7 Formulate complete state space representation of series motor from Fig. 6.16 by measuring voltage and current of the load at the output.

P6.8 Formulate the state space representation from the circuits with fields in Fig. 6.23.

P6.9 There are three resistive elements in the Bridge circuit given in Fig. 6.17. Find out efforts or flow measurements in each; give a matrix of mixed field form accordingly.

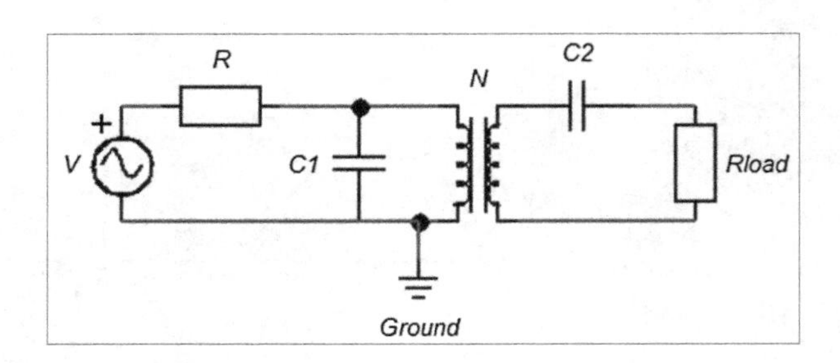

Fig. 6.23 Circuit with induction transformer

P6.10 If capacitors in problem P6.6 are replaced with resistors and voltage across each resistor, and flow in each branch needs to be measured. Provide a generalized matrix to give the measurement and specify how many elements in vector to measure effort and how many elements for measuring flow in a mixed form causality assignment.

P6.11 Simplify the Bond Graph given in P6.1 by eliminating the bond 5 (from both parts).

References

1. Karnopp, Dean C., Donald L. Margolis, and Ronald C. Rosenberg. 2012. *System Dynamics—Modeling and Simulation of Mechatronics Systems*, 5th ed. Hoboken: Wiley.
2. 20-Sim Software. www.20sim.com.
3. Zimmer, Dirk., François E. Cellier. 2006. The Modelica multi-bond graph library. In *Proceedings of the 5th International Modelica Conference, Vienna, Austria*, 4–6 September 2006, 559–568.
4. Craig, K., P. Voglewede. 2010. Multidisciplinary Engineering Systems Graduate Education: Master of Engineering in Mechatronics. In *Transforming Engineering Education: Creating Interdisciplinary Skills for Complex Global Environments, IEEE, Dublin*, 6–9 April 2010, 1–14.
5. Cetinkunt, Sabri. 2006. *Mechatronics*, 1st ed. Somerset: Wiley.

Chapter 7
Simulation and Analysis of State Space Systems

State Space formulation of a system belongs to vector space either in real or complex vector fields. The solution of second-order or higher order state space requires concept of linear (matrix) algebra for understanding. First-order state space system solves by simple theorems of calculus and algebra. The first-order equation may be homogenous or non-homogenous and both require the standard solution. The first-order linear time invariant state space is

$$\dot{x}(t) = ax(t) + bu(t) \tag{7.1}$$

This equation is non-homogenous due to the term $bu(t)$, so the solution requires computation of the homogenous and particular solution given as

$$x(t) = x_h(t) + x_p(t) \tag{7.2}$$

The homogenous solution $x_h(t)$ obtains by initial conditions and system is solved if $u(t) = 0$ and general homogenous solution of Eq. (7.1) is given as

$$x_h(t) = \mathrm{e}^{a \cdot t} \cdot x(0) \tag{7.3}$$

The particular solution $x_p(t)$ is solved by calculus due to function $u(t)$ in the equation and total solution of Eq. (7.1) is given as

$$x(t) = \mathrm{e}^{a \cdot t} \cdot x(0) + \mathrm{e}^{at} \cdot \int_0^t \mathrm{e}^{-a \cdot \tau} \cdot b \cdot u(\tau) d\tau \tag{7.4}$$

We can see from Eq. (7.2) that at $t = 0$ system starts responding from homogenous solution and if $u(t) = 0$ then only the homogenous solution contributes to the final shape of the response. If the system is relaxed i.e., $x(0) = 0$ then the system is only excited by the given input $u(t)$. A system excites to either non-zero initial conditions

A.M. Mughal, *Real Time Modeling, Simulation and Control of Dynamical Systems*,
DOI 10.1007/978-3-319-33906-1_7

or given input, otherwise the system remains at rest. These two excitation sources are known as ***free response*** which are due to initial conditions only and ***forced response*** due to given input to the system. The output state space response of the first-order system is

$$y(t) = c \cdot \left\{ e^{a \cdot t} \cdot x(0) + \int_0^t e^{a \cdot (t-\tau)} \cdot b \cdot u(\tau) d\tau \right\} + d \cdot u(t) \tag{7.5}$$

The same concept applies to higher order state space system

$$\begin{aligned} \dot{\vec{x}}(t) &= A \cdot \vec{x}(t) + B \cdot \vec{u}(t) \\ \vec{y}(t) &= C \cdot \vec{x}(t) + D \cdot \vec{u}(t) \end{aligned} \tag{7.6}$$

This equation is solved by linear algebra and the general form of the solution in vector field is

$$\vec{x}(t) = e^{A \cdot t} \cdot \vec{x}(0) + e^{A \cdot t} \cdot \int_0^t e^{-A \cdot \tau} \cdot B \cdot \vec{u}(\tau) d\tau \tag{7.7}$$

In this case A, B are matrices of $n \times n$ and $n \times p$ orders respectively, where $\vec{x}(t)$, $\vec{u}(\tau)$ are vectors of $n \times 1$ and $p \times 1$ orders respectively. The output response solution is

$$\vec{y}(t) = C \cdot \left\{ e^{A \cdot t} \cdot \vec{x}(0) + \int_0^t e^{A \cdot (t-\tau)} \cdot B \cdot \vec{u}(\tau) d\tau \right\} + D \cdot \vec{u}(t) \tag{7.8}$$

In the output response, $\vec{y}(t)$ is a vector of $q \times 1$ order and C, D are matrices of $q \times n$ and $q \times p$ orders respectively. The general formulation of state space equation of any order solves by integrating the parameters with given input over the time. The simulation diagram of the system in Fig. 7.1, illustrates the output Eq. (7.8) solving state space formulation of Eq. (7.6).

The figure shows the real-time simulation diagram in MATLAB/Simulink of state space equations; an integrator integrates the vector $\dot{\vec{x}}(t)$ during the simulation. This vector $\dot{\vec{x}}(t)$ is composed of two terms, first is the multiplication of state vector A with output of integrator which is the state vector $\vec{x}(t)$ and the term is input vector $\vec{u}(t)$ multiplied with input gain matrix B. Similarly, the output vector is composed of two terms, first is the multiplication of output gain matrix C to state vector $\vec{x}(t)$ and a feedforward component from input $\vec{u}(t)$ by multiplying feedforward (input–output) matrix D. If the system is of first order then all matrices and vectors are

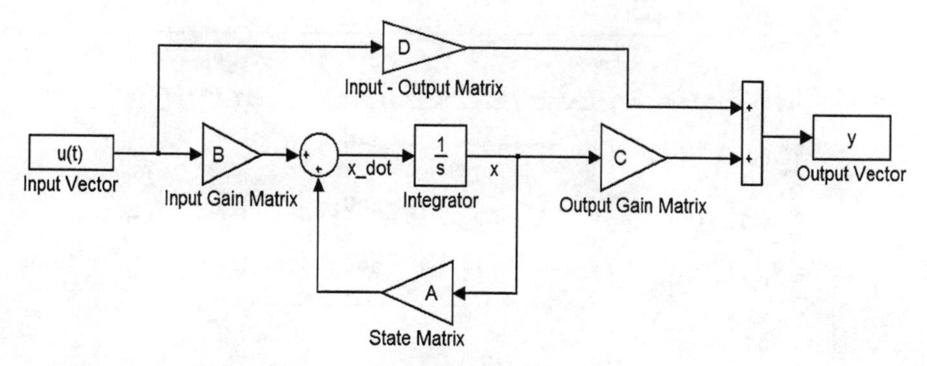

Fig. 7.1 Simulation diagram of a state space system

treated as 1×1 elements to represent Eq. (7.1). This system can also be simulated as a transfer function approach, which is only for a relaxed system and for which initial conditions cannot be specified in simulation. In this block diagram, the integrator takes initial conditions as a vector and starts processing from these values to determine the total response of the system.

7.1 Free Response of First-Order System

The free response of a system is obtained with zero input and exciting through initial conditions only. The homogenous solution of an nth order differential equation either in algebraic formulation or in state space formulation provides the free response of the system. The free response of the first-order system is given as $x(t) = \mathrm{e}^{a \cdot t} \cdot x(0)$ and with output it is given as

$$y(t) = c \cdot \mathrm{e}^{a \cdot t} \cdot x(0) \tag{7.9}$$

Let us consider a system

$$\begin{aligned} \dot{x}(t) &= a\,x(t) + u(t) \\ y(t) &= x(t) \end{aligned} \tag{7.10}$$

The free response of the system is given as

$$y(t) = \mathrm{e}^{a \cdot t} \cdot x(0) \tag{7.11}$$

If the initial condition is zero then the output remains silent for all the time but if there is non-zero initial condition then the output will start from that value and either increase or decrease exponentially depending upon the sign of parameter a. In a first-order system, this parameter a or matrix $[a]$ constitutes the only eigenvalue

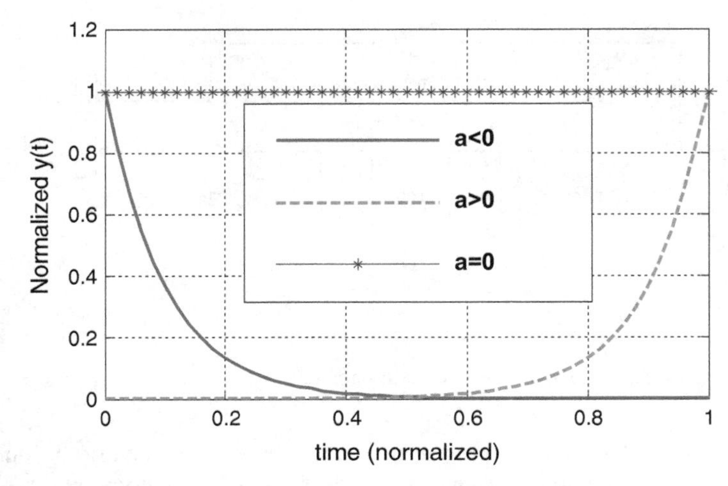

Fig. 7.2 Normalized responses of a first-order system with change in parameter {eigenvalue} a

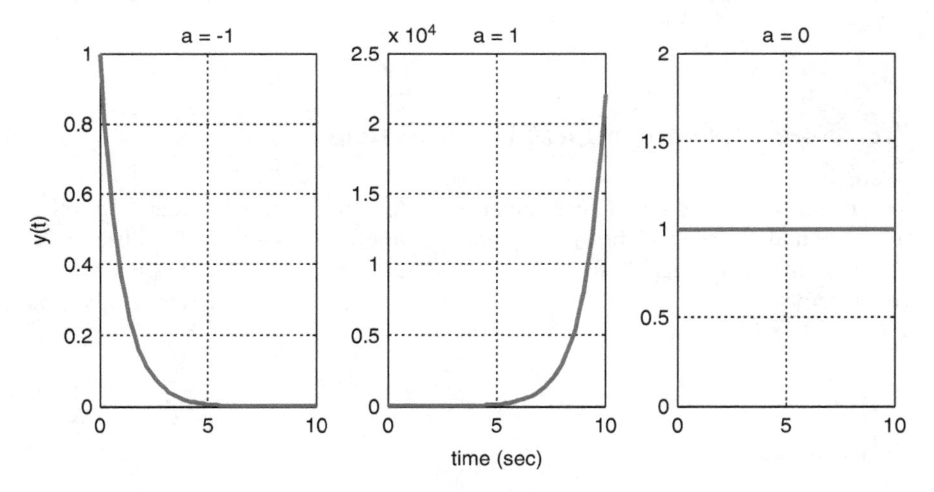

Fig. 7.3 Amplitude of responses of a first-order system in 10 s by varying eigenvalue

of the system. If $a > 0$ then the response will increase exponentially and if $a < 0$ then the response will decrease exponentially. The response will remain constant if $a = 0$ at initial condition. In a first-order case we can simply observe that output gain matrix C just amplifies or attenuates the response. Free responses in Fig. 7.2 show the dependency on a, which is the only eigenvalue of the first-order system. The output is normalized to maximum value in order to show the comparison of responses for exponential growth and decay for the same eigenvalue with sign inversion. The responses for three cases are also shown in Fig. 7.3 to show the level of amplitude obtained in 10 s time.

7.2 Eigenvalues of Higher Order System

An nth order system will have n eigenvalues; the free response of such a system depends upon all the eigenvalues. A system matrix can be constituted such that all eigenvalues appear explicitly in state matrix, but generally this is not the case. The eigenvalues are also called poles of the system, which are determined from the denominator of transfer function. A transfer function is a realization of the relaxed state space system in frequency domain as we already discussed in Chap. 1. A system with p inputs and q outputs has $p \times q$ transfer functions from each output to each input.

$$\vec{G}(s) = C \cdot [s \cdot I - A]^{-1} \cdot B + D \tag{7.12}$$

Each transfer function shows the relationship between a single input to a single output represented as

$$G(s) = \frac{N(s)}{D(s)} = \frac{\beta_m \cdot s^m + \beta_{m-1} \cdot s^{m-1} \ldots \beta_2 \cdot s^2 + \beta_1 \cdot s + \beta_0}{s^n + \alpha_{n-1} \cdot s^{n-1} \ldots \alpha_2 \cdot s^2 + \alpha_1 \cdot s + \alpha_0} \tag{7.13}$$

The denominator $D(s) = s^n + \alpha_{n-1} \cdot s^{n-1} \ldots \alpha_2 \cdot s^2 + \alpha_1 \cdot s + \alpha_0$ is the nth order polynomial with n solutions for $D(s) = 0$ and each solution is a pole of a system, which are all eigenvalues of state matrix $[A]$. The equation $D(s) = 0$ is known as the characteristic equation of the system and also given by the computing determinant as follows

$$D(s) = |s \cdot I - A| = 0 \tag{7.14}$$

The response of each eigenvalue is similar to the response of the first-order system. If an eigenvalue is less than zero then it contributes a decaying response and if an eigenvalue is greater than zero then it grows exponentially. In higher order real systems, complex eigenvalues may appear in a conjugate pair. If $D(s) = 0$ has any complex root then its conjugate pair is also a root of the polynomial $D(s)$. This is due to the quadratic solution of the polynomial; a third-order polynomial may have three real roots or a real root and two complex roots (in which both must be conjugate to each other). In a matrix notation, eigenvalues may appear in diagonal of $[A]$ matrix if the matrix is diagonal or upper triangular but for real systems, complex eigenvalues do not appear in complex form in real matrices. Finding roots of the polynomial $D(s)$ or obtaining eigenvalues through matrix decomposition presents the complex eigenvalues. The response of nth order system realizes as the combination of n first-order systems if all poles are real, whereas we can also observe the response of nth order system as a combination of second-order system if roots are complex conjugate pair with the remaining first-order system. A transfer function given in Eq. (7.13) can also be written as

$$G(s) = K\frac{\prod_{j=1}^{m}(s - s_j)}{\prod_{i=1}^{p}(s - s_i)}. \tag{7.15}$$

Where there are m zeros that are roots of $N(s)$ and n poles that are roots of $D(s)$. Each real root has the simple form $s = s_i$ where s_i is a real number. A complex root has a form

$$s_i = \sigma + j \cdot \omega \tag{7.16a}$$

And conjugate pair is also a root i.e.,

$$s_i^* = \sigma - j \cdot \omega \tag{7.16b}$$

The combination of conjugate pairs of roots forms a section of polynomial as

$$D_2(s) = (s - s_i) \cdot (s - s_i^*) = (s - \sigma - j \cdot \omega) \cdot (s - \sigma + j \cdot \omega) \tag{7.17}$$

Which yield as

$$D_2(s) = s^2 + a \cdot s + b \tag{7.18}$$

For example a third-order system can be given as

$$G(s) = \frac{k \cdot \prod_{j=1}^{m}(s - s_j)}{(s - s_1) \cdot D_2(s)} \quad m \leq 3 \tag{7.19}$$

It is not possible to have a real system with complex coefficients in either denominator or numerator as it cannot be implemented with real devices. So poles or zeros always exist in conjugate pairs which lead to polynomials with real coefficients.

7.3 Free Response of a Second-Order System

A second-order system consists of two poles and the location of these poles in complex plane "s" describes the behavior of response. A solution of higher order system in Eq. (7.8) may appear as a combination of two linear solutions of the first-order system

$$y(t) = y_1(t) + y_2(t) \tag{7.20}$$

Note here $y_1(t)$ and $y_2(t)$ are two first-order solutions due to two eigenvalues (poles) of the second-order system λ_1 and λ_2. The solution will be

$$y_1(t) = c_1 \cdot e^{\lambda_1 \cdot t} \quad y_2(t) = c_2 \cdot e^{\lambda_2 \cdot t} \tag{7.21}$$

Where the second-order system of characteristic equation $D_2(s) = s^2 + a \cdot s + b = 0$ gives the solution of λ_1 and λ_2

$$\lambda_1 = \frac{-a + \sqrt{a^2 - 4b}}{2} \qquad \lambda_2 = \frac{-a - \sqrt{a^2 - 4b}}{2} \tag{7.22}$$

The values or a and b forms two types of responses, undamped and damped responses. These are also characterized as undamped oscillation or damped oscillation of the second-order system.

7.3.1 Undamped Systems

If $a = 0$ then the system is undamped and two eigenvalues are conjugate pairs given as

$$\lambda_{1,2} = \pm\sqrt{-b} = \pm j\omega \tag{7.23}$$

In this case we refer $\omega = \sqrt{b}$ as undamped natural frequency of the system. The solution of the system is given as

$$y(t) = c_1 \cdot e^{j\omega \cdot t} + c_2 \cdot e^{-j\omega \cdot t} \tag{7.24}$$

By using Euler identities this solution can also be represented as

$$y(t) = k_1 \cdot \cos(\omega \cdot t) + k_2 \cdot \sin(\omega \cdot t) \tag{7.25}$$

Example 7.1 Undamped Oscillations:

Find the transfer function and solution of the following state space system and plot the free response of the system with different initial conditions

$$\dot{\vec{x}}(t) = \begin{bmatrix} 0 & -1 \\ 1 & 0 \end{bmatrix} \cdot \vec{x}(t) + \begin{bmatrix} 1 \\ 1 \end{bmatrix} \cdot u(t)$$

$$y(t) = \begin{bmatrix} 1 & 1 \end{bmatrix} \cdot \vec{x}(t)$$

$$\vec{x}_{01}(0) = \begin{bmatrix} 1 \\ 0 \end{bmatrix} \quad \vec{x}_{02}(0) = \begin{bmatrix} 0 \\ 1 \end{bmatrix} \quad \vec{x}_{03}(0) = \begin{bmatrix} 1 \\ 1 \end{bmatrix}$$

Solution The transfer function is obtained from MATLAB by using the following command

```
A=[0 -1;1 0]
B=[1;1]
C=[1 1]
```

```
D=0
[Num,Den]=ss2tf(A,B,C,D)
```

$$G(s) = \frac{2s}{s^2 + 1} \tag{7.26}$$

The system has two eigenvalues which can be found by either command `roots (Den)` or `eig(A)` to obtain roots of characteristic equation or eigenvalues of $[A]$ computed as

```
0 + 1i
0 - 1i
```

Now we can determine solution as

$$y(t) = c_1 \cdot e^{j \cdot t} + c_2 \cdot e^{-j \cdot t} \quad \text{or} \quad y(t) = k_1 \cdot \cos(t) + k_2 \cdot \sin(t) \tag{7.27}$$

We can use initial conditions to find the values of constants in either of two solutions. Algebraically in this problem we plug initial conditions to the equation $y(t)$ and $\dot{y}(t)$ or we can use free response simulation as $\vec{y}(t) = e^{A \cdot t} \cdot \vec{x}(0)$. We can also directly simulate the system in SIMULINK given in Fig. 7.1 with zero input and obtain solutions for different initial conditions given in Fig. 7.4

We can see that responses two $\vec{x}_{01}$ and $\vec{x}_{02}$ are 90° out of phase because either $\cos(t)$ or $\sin(t)$ will remain in a solution. Using initial conditions $\vec{x}_{03}$ both terms share their parts in the solution of free response. All three initial conditions yield a sinusoidal signal in response without any amplification or attenuation over time, this is known as undamped oscillation of the system. The undamped oscillatory response only appears when real parts of second-order poles are zero or we can say eigenvalues (poles) lie on the imaginary axis of s-plane.

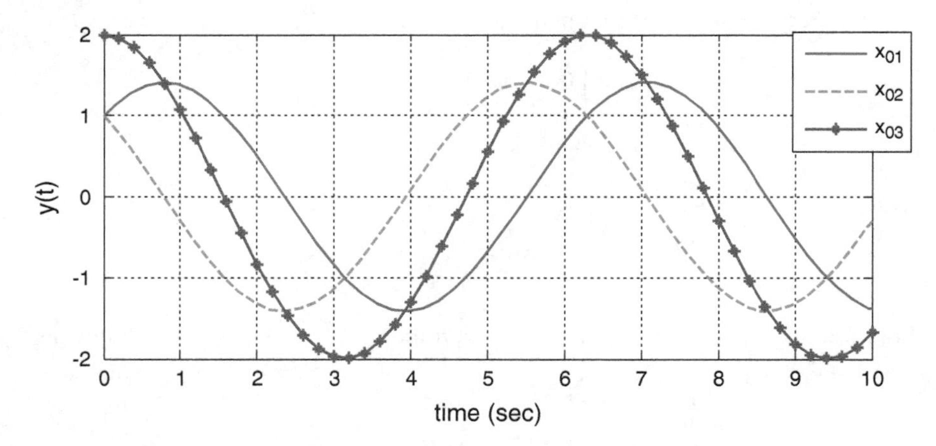

Fig. 7.4 Free response of undamped system with different initial conditions

7.3.2 *Damped Systems*

A second-order system of characteristic equation $s^2 + a \cdot s + b = 0$ has two roots given as

$$\lambda_1 = \frac{-a + \sqrt{a^2 - 4b}}{2} \qquad \lambda_2 = \frac{-a - \sqrt{a^2 - 4b}}{2}$$

and if parameter $a \neq 0$ then the system has damped oscillations based upon the value of a. Let us now consider

$$\lambda_{1,2} = \sigma \pm j\omega = \frac{-a}{2} \pm j\frac{\sqrt{4b - a^2}}{2} \tag{7.28}$$

The term $\sigma = {}^{-a}/_{2}$ is known as damping frequency and also referred as attenuation ratio. The term ω is the frequency of the system which is measured in $^{\text{rad}}/_{\text{s}}$ and related with time period T as $\omega = {}^{2\pi}/_{T}$ and frequency $\omega = 2\pi f$. The characteristic equations of the second-order system often realize with two parameters of damping ratio ζ (zeta) and natural frequency ω_n.

$$D_2(s) = s^2 + a \cdot s + b = s^2 + 2\,\zeta\omega_n \cdot s + {\omega_n}^2 = 0 \tag{7.29}$$

Here we can see that $a = 2\,\zeta\omega_n$ and $b = {\omega_n}^2$ as equivalence between two representation of the same equations. The roots of new representations are

$$\lambda_{1,\,2} = -\zeta\omega_n \pm j\omega_n\sqrt{1 - \zeta^2} \tag{7.30}$$

Here we can see that $\sigma = -\zeta\omega_n$ and $\omega = \omega_n\sqrt{1 - \zeta^2}$ as real and imaginary parts of complex roots. Both cases show the damped oscillation and with negative real parts of complex roots, response always attenuates with time. If the damping ratio is zero i.e., $\zeta = 0$ then the system is undamped and behaves exactly like Eq. (7.27); for $\zeta > 0$ there are three types of damped oscillations.

7.3.2.1 Case I: Overdamped System

A system is underdamped with two distinct real roots because there are no imaginary parts in the complex roots. This case arises when

$a^2 - 4b > 0$ or $a^2 > 4b$ and $\zeta^2 > 1$ or $(\zeta > 1)$.

The system will have two distinct real roots given in Eq. (7.22)

$$\lambda_{1,\,2} = \frac{-a \pm \sqrt{a^2 - 4b}}{2}$$

The solution of the system is

$$y(t) = c_1 \cdot e^{\lambda_1 \cdot t} + c_2 \cdot e^{\lambda_2 \cdot t} \tag{7.31}$$

In this case we may get both positive, both negative, or a positive and negative root but for overdamped systems both roots are negative. With negative roots as $t \to \infty$, $y(t) \to 0$ so quickly that it does not allow the system to oscillate and there is no damped natural frequency in the system. If roots are very small as compared to sluggish model parameters (a, b), there may be overshoot or undershoot depending upon the initial conditions of the system.

Example 7.2 Overdamped Oscillations

Find the free response of the following system with given initial conditions and explain why this is an overdamped system.

$$\dot{\vec{x}}(t) = \begin{bmatrix} -1 & 0 \\ 0 & -2 \end{bmatrix} \cdot \vec{x}(t) + \begin{bmatrix} 1 \\ 1 \end{bmatrix} \cdot u(t)$$
$$y(t) = \begin{bmatrix} 1 & 1 \end{bmatrix} \cdot \vec{x}(t)$$

$$\vec{x}_{01}(0) = \begin{bmatrix} 1 \\ 1 \end{bmatrix} \quad \vec{x}_{02}(0) = \begin{bmatrix} -1 \\ -1 \end{bmatrix} \quad \vec{x}_{03}(0) = \begin{bmatrix} -1 \\ 1 \end{bmatrix} \quad \vec{x}_{04}(0) = \begin{bmatrix} 1 \\ -1 \end{bmatrix}$$

Solution The system transfer function calculates as follows by using symbolic math toolbox

```
A=[-1 0;0 -2];
B=[1; 1];
C=[1 1];
D=0;
syms s
G=C*inv(s*eye(2,2)-A)*B
collect(G)
```

The transfer function is

$$G(s) = \frac{2s + 3}{s^2 + 3s + 2} \tag{7.32}$$

The characteristic equation $D(s) = s^2 + 3s + 2$ gives us two roots at $s = -1$ & $s = -2$, which are also eigenvalues of the A matrix. These both are distinct and we also know that in this case $a^2 > 4b$ $(9 > 8)$ so the system will be overdamped. We now simulate the system as given in Fig. 7.4 to get the free response at 4 different initial conditions. It is evident in the figure that if both initial conditions are positive or negative then in this case systems damped out without any oscillations. In case the initial conditions are mixed then the system has either undershoot for $\vec{x}_{03}(0)$ or overshoot for $\vec{x}_{04}(0)$. This also depends upon which eigenvalue (pole) dominates over others in response; higher magnitude eigenvalues are dominant over lesser amplitude eigenvalues (Fig. 7.5).

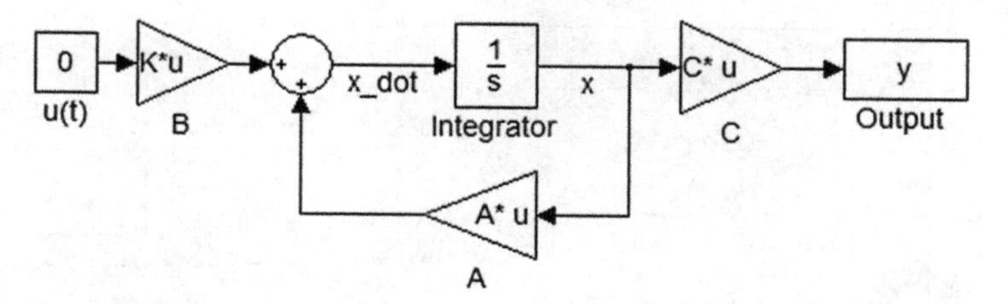

Fig. 7.5 MATLAB/Simulink diagram for simulation of free response

Damping analysis of characteristic equations also tells us that $\omega_n = \sqrt{2}$ & $\zeta = \frac{3}{2\sqrt{2}} > 1$, which also describes it as overdamped system.

7.3.2.2 Case II: Critically Damped System

A second-order system is said to be critically damped if it has two real double roots and parametrically $a^2 - 4b = 0$ or $a^2 = 4b$. This case deals with a class of second-order system with unity damping ratio $\zeta^2 = 1$ or $(\zeta = 1)$. The general solution to this type of system with both same roots $\lambda = -a/2$ is given as

$$y(t) = c_1 \cdot e^{\lambda \cdot t} + c_2 \cdot t \cdot e^{\lambda \cdot t} \tag{7.33}$$

This system will lead to a very small constant value because of term $t \cdot e^{-a/2 \cdot t}$ and non-zero constant c_2. As $t \to \infty$, $e^{-a/2 \cdot t} \to 0$ but a term $t \cdot e^{-a/2 \cdot t}$ makes an indeterminate form where a limit can be found closer to zero.

Example 7.3 Critically Damped Oscillations

Find the free response of the following system with given initial conditions and explain why this is a critically damped system.

$$\dot{\vec{x}}(t) = \begin{bmatrix} -1 & 1 \\ 0 & -1 \end{bmatrix} \cdot \vec{x}(t) + \begin{bmatrix} 1 \\ 1 \end{bmatrix} \cdot u(t)$$
$$y(t) = \begin{bmatrix} 1 & 1 \end{bmatrix} \cdot \vec{x}(t)$$

$$\vec{x}_{01}(0) = \begin{bmatrix} 1 \\ 1 \end{bmatrix} \quad \vec{x}_{02}(0) = \begin{bmatrix} -1 \\ -1 \end{bmatrix} \quad \vec{x}_{03}(0) = \begin{bmatrix} -1 \\ 1 \end{bmatrix} \quad \vec{x}_{04}(0) = \begin{bmatrix} 1 \\ -1 \end{bmatrix}$$

Solution There are two eigenvalues/poles of the system at $s = -1, -1$. The transfer function of the system is given as

$$G(s) = \frac{2s + 3}{s^2 + 2s + 1} = \frac{2\left(s + \frac{3}{2}\right)}{(s + 1)^2} \tag{7.34}$$

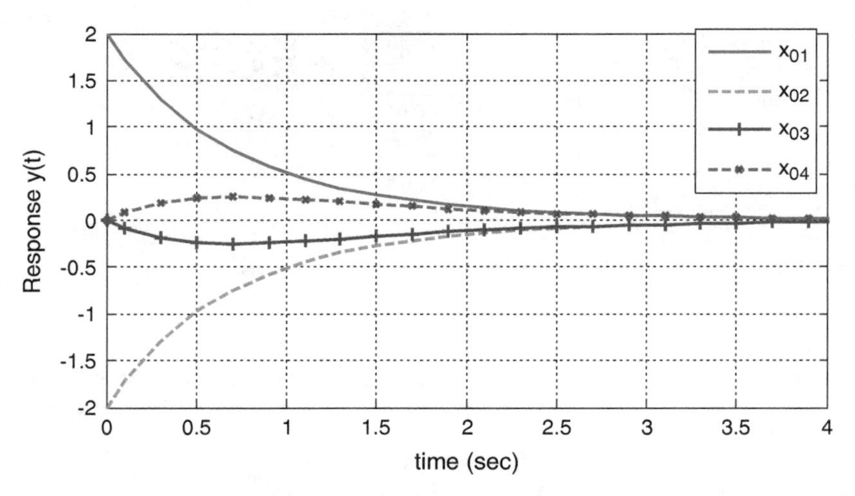

Fig. 7.6 Free response of a overdamped system

The zero-poles-gain form of transfer function also shows that there is one zero at $s = -\frac{3}{2}$ and two poles at $s = -1, \quad -1$. This form can be obtained in MATLAB by the following command:

```
sys
sys=ss(A,B,C,D)
sys1=zpk(sys)
```

Responses with different initial conditions obtained through simulation of block diagram are given in Fig. 7.4.

It is evident from characteristic equations that $a^2 = 4b$ and so there is no imaginary part in roots. From damping analysis we can also see that natural frequency $\omega_n = 1$ as well as $\zeta^2 = 1$ or $(\zeta = 1)$. The response is similar in shape but slower in response time, this response will only be faster than the overdamped system if the magnitude of double eigenvalue (double root) is equal or greater than the maximum magnitude of eigenvalues from overdamped system (Fig. 7.6).

7.3.2.3 Case III: Underdamped Systems

A second-order system may have both complex roots, as we consider in Eq. (7.26). The system will oscillate like undamped systems but the difference is the magnitude of damping will decrease with time due to a damping ratio ζ. The case arises when $(a^2 - 4b) < 0$ or $\zeta^2 < 1$ or $(0 < \zeta < 1)$. The eigenvalues are represented as

$$\lambda_{1,2} = -\zeta\omega_n \pm j\omega_n\sqrt{1-\zeta^2} = \frac{-a}{2} \pm j\frac{\sqrt{4b-a^2}}{2} = \sigma \pm j\omega \tag{7.35}$$

Now we can find a general solution as

$$y(t) = c_1 \cdot e^{(\sigma+j\omega)\cdot t} + c_2 \cdot e^{(\sigma-j\omega)\cdot t} \tag{7.36}$$

There are other representations of this solution as

$$\begin{aligned} y(t) &= e^{\sigma \cdot t}\{k_1 \cos(\omega t) + k_2 \sin(\omega t)\} \\ &= K \cdot e^{\sigma \cdot t} \cdot \cos(\omega t - \delta) \\ K &= \sqrt{{k_1}^2 + {k_2}^2} \quad \delta = \tan^{-1}\left({}^{k_2}/_{k_1}\right) \end{aligned} \tag{7.37}$$

In each of the solutions there are two constants $\{(c_1,\ c_2),\ (k_1,\ k_2),\ (K,\ \delta),\}$ which can be found from the given initial conditions. $(c_1,\ c_2)$ and $(k_1,\ k_2)$ are simple amplitude gain constants whereas K and δ are amplitude and phase difference (leading or lagging depending upon the sign of phase difference) respectively. For underdamped systems as $t \to \infty$, the response is $y(t) \to 0$ after oscillating in sinusoidal fashion.

Example 7.4 Underdamped Oscillations

Find the free response of the following system with given initial conditions and explain why this is an underdamped system.

$$\begin{aligned} \dot{\vec{x}}(t) &= \begin{bmatrix} -1 & 0 \\ 0 & -2 \end{bmatrix} \cdot \vec{x}(t) + \begin{bmatrix} 1 \\ 1 \end{bmatrix} \cdot u(t) \\ y(t) &= \begin{bmatrix} 1 & 1 \end{bmatrix} \cdot \vec{x}(t) \end{aligned}.$$

$$\vec{x}_{01}(0) = \begin{bmatrix} 1 \\ 1 \end{bmatrix} \quad \vec{x}_{02}(0) = \begin{bmatrix} -1 \\ -1 \end{bmatrix} \quad \vec{x}_{03}(0) = \begin{bmatrix} -1 \\ 1 \end{bmatrix} \quad \vec{x}_{04}(0) = \begin{bmatrix} 1 \\ -1 \end{bmatrix}$$

Solution Figure 7.7 shows that initial conditions vary in amplitude of response as well as phase with respect to zero. These responses die down to zero as $t \to \infty$ while

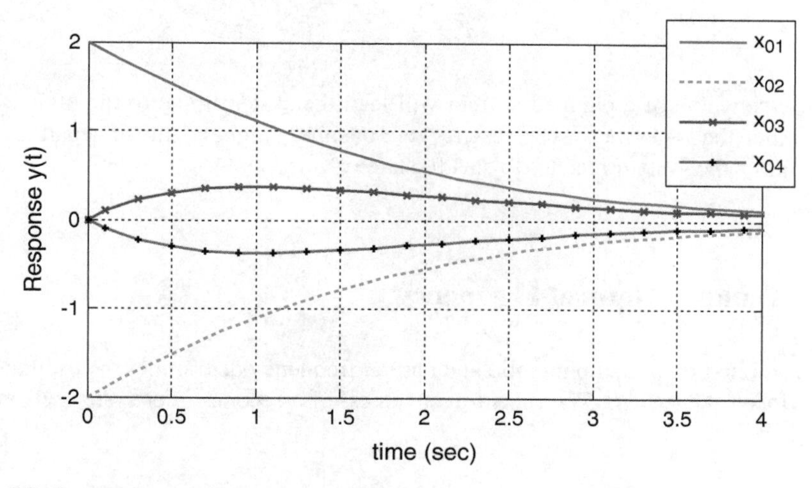

Fig. 7.7 Free response of an critically damped system

oscillating at frequency ω. The first and second initial conditions are both positive or negative so these constitute higher initial magnitude of response while 180° out of phase from each other, similarly third and fourth initial conditions have mixed signs and so these have lesser amplitude but also 180° out of phase with each other.

7.4 Relationship Between Eigenvalue and Time

General solutions of undamped and damped oscillations can be written as

$$y(t) = c_1 \cdot \mathrm{e}^{(\sigma+j\omega)\cdot t} + c_2 \cdot \mathrm{e}^{(\sigma-j\omega)\cdot t} \tag{7.38}$$

This is a general solution and different cases makes it undamped, overdamped, critically damped, and underdamped systems. In the general solution of undamped system, we see that there is no exponential term with damping ratio. The exponential term with complex power resolves as sinusoidal terms with constant amplitude. The value of imaginary term i.e., ω is the frequency of oscillation. In case of overdamped and critically damped systems there is no frequency of oscillation but there is an exponential term for each eigenvalue. For underdamped systems there is an exponential term for damping and sinusoidal terms for oscillations. Like an undamped system, ω is the oscillation frequency in underdamped system but these oscillations will decrease exponentially because of an exponential term $\mathrm{e}^{\sigma\cdot t}$. The exponential decaying can be seen as an envelope of underdamped system to depict change in magnitude in each cycle. In a damped system, the term $\sigma = \{\sigma,\ \lambda,\ \lambda_1,\ \lambda_2\}$ for different damped systems is always negative and so the system will approach zero exponentially as $t \to \infty$. The inverse of this term is also known as time constant τ, usually in electrical circuits where each multiple of τ represents an order decrease of magnitude from initial condition at zero.

$$\tau = \frac{1}{\sigma} \text{ or } \frac{1}{\lambda} \tag{7.39}$$

Each eigenvalue of a damped system will lead the system to zero; the higher the magnitude the faster the system reaches zero or simply higher value of negative real part, faster the system reaches to steady-state or zero.

7.5 Damped Natural Frequency

The combination of damping ratio and natural frequency determines the oscillatory behavior of the system. The roots (eigenvalues) of a second-order system given in Eq. (7.27) are

$$\lambda_{1,2} = -\zeta\omega_n \pm j\omega_d \tag{7.40}$$

Where $\omega_d = \omega_n\sqrt{1-\zeta^2}$ is known as damped natural frequency of the system and so a quadratic equation for the second-order system can be given as

$$\left(\frac{\omega_d}{\omega_n}\right)^2 + \zeta^2 = 1 \tag{7.41}$$

We can observe that as $\zeta \rightarrow 1$ then $\omega_d \rightarrow 0$ which means an undamped system and as $\zeta \ll 1$ then the $\omega_d \rightarrow \omega_n$. Let us consider a transfer function

$$G(s) = \frac{N(s)}{s^2 + 2\zeta\omega_n + {\omega_n}^2} \tag{7.42}$$

We select $\omega_n = 1$ and vary $\zeta \rightarrow \{0, \quad 1\}$ with steps of 0.1 to get a response by converting into a state space system. The amplitude of *y*(*t*) damped down with time while oscillating at natural frequency and decrease in amplitude corresponds to the damping ratio. The damped natural frequency depends upon both ω_n and ζ and so the transfer function of a system may have different behavior as corresponding to different damping ratios. Figure 7.8 shows the Bode plot of transfer function given in Eq. (7.41). The Bode plot shows the change in magnitude of the transfer function and phase of transfer function with respect to frequency. The first plot shows the magnitude of the transfer function $|G(j\omega)|$ and the second plot shows the phase $\angle(G(j\omega))$ of the transfer function $G(s)$ at $s = j\omega$. We plot the Bode diagram of state space (or transfer function) of the system of Eq. (7.41) with $N(s) = 1$. We can

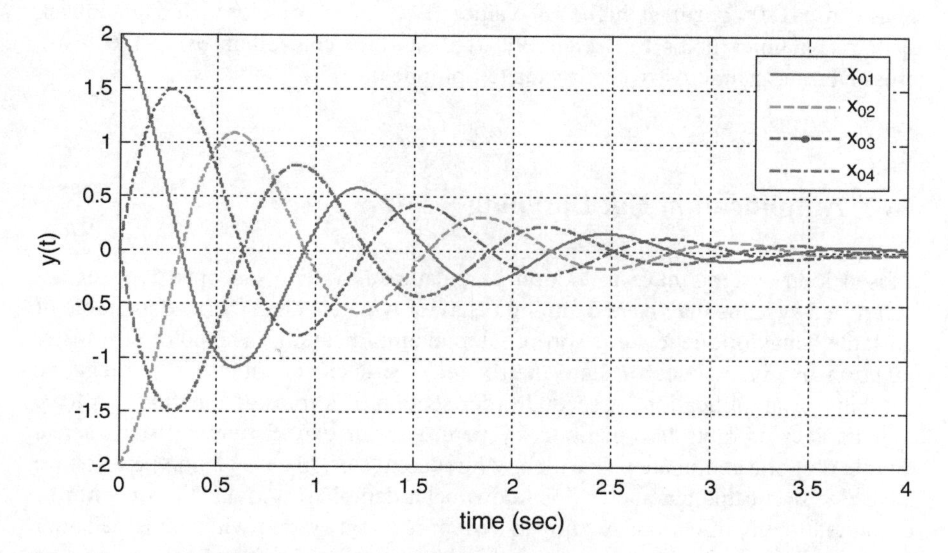

Fig. 7.8 Free response of an under-damped system

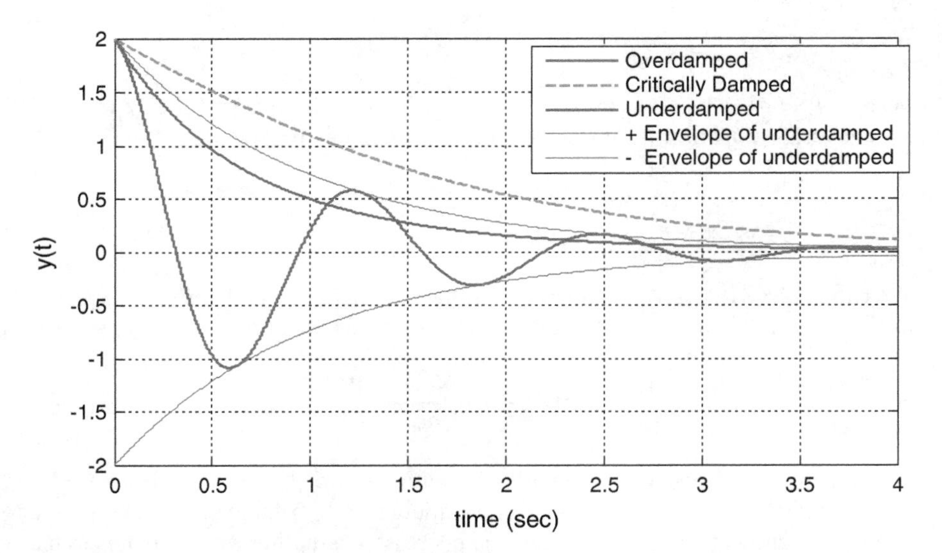

Fig. 7.9 Free response of overdamped, critically damped and under damped {with envelopes on starting from positive or negative initial conditions} systems

observe that at $\omega = \omega_n$, there is a peak and the amplitude of this peak depends upon the damping ratio ζ; the higher the ζ, the lower the amplitude of $|G(j\omega)|$. This corresponds to Fig. 7.9 that higher ζ will underdamp the system and lower ζ will tend towards overdamping. In other part of the Bode diagram which shows the relationship of $\angle(G(j\omega))$ with frequency, a steep shift from $0°$ to $-180°$ occurs at ω_n for $\zeta = 0$ and as $\zeta \to 1$, the curve of $\angle(G(j\omega))$ becomes more smooth and the phase of $-180°$ occur at higher ω values. Bode plots of closed loop feedback system dynamics helps to design PID (or lead-lag) controllers using the phase crossover and gain crossover margins (amplitudes).

7.6 Amplification Not Damping

Closed loop systems have to exhibit as a damped system, but open loop or not controlled systems may have damping behavior which is characterized as stable or unstable behavior due to some sort of internal amplifications. The poles of transfer function or eigenvalues of state matrix may be such that it produces negative damping or amplification. A second-order section of a transfer function can have four choices of roots like damping system; a set of two distinctive roots, a real double root or a conjugate pair with or without real part. In case of amplification we have $\zeta < 0$ but this parameter ζ is known as a damping ratio and so it cannot be attributed to amplification. A damped system is a real system with stable behavior and an amplified system is a real system with unstable behavior so these terms are not used for describing amplification systems. The behavior of these types of systems remains similar but instead of damping down, the amplitude of response will be amplified. Now let us consider a second-order system with different combination of poles.

$$G(s) = \frac{N(s)}{D(s)} = \frac{N(s)}{s^2 + as + b} = \frac{N(s)}{(s - \sigma + j\omega) \cdot (s - \sigma - j\omega)} \tag{7.43}$$

The solution of the system will be given in Eq. (7.21) as follows

$$y(t) = c_1 \cdot \mathrm{e}^{\lambda_1 \cdot t} + c_2 \cdot \mathrm{e}^{\lambda_2 \cdot t} \tag{7.44}$$

Where the second-order system of characteristic equation $D_2(s) = s^2 + a \cdot s + b = 0$ gives the solution of λ_1 and λ_2

$$\lambda_{1,2} = \sigma \pm j\omega = \frac{-a}{2} \pm j\frac{\sqrt{4b - a^2}}{2}$$

A system will be undamped exactly like given in Eq. (7.25) and Fig. 7.10 if there is no damping terms ζ or $a = 0$ or $\sigma = 0$. The response in Fig. 7.10 is classified as undamped and unamplified response. Now there are three other choices of roots with $\sigma \neq 0$. A system has two distinct roots of $a^2 > 4b$, a system may have a real

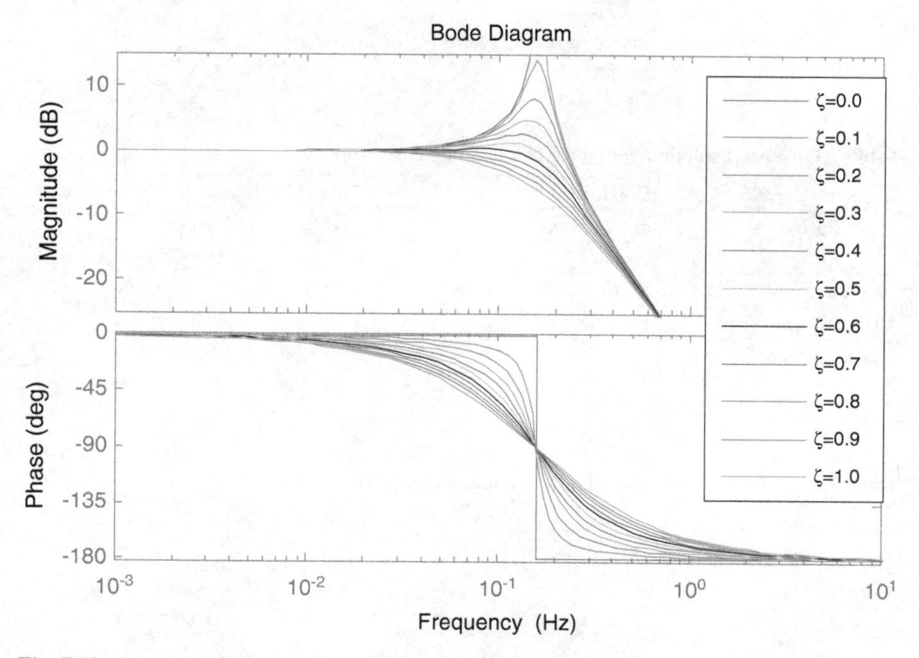

Fig. 7.10 Bode plot of a second order system

double root if $a^2 = 4b$ i.e., no imaginary part and a system may have a conjugate pair if $a^2 < 4b$.

The response shown in Fig. 7.11 of Table 7.2 is similar to responses of Table 7.1 but the difference is amplitude of response increases with time whereas, in a damped system it decreases with time.

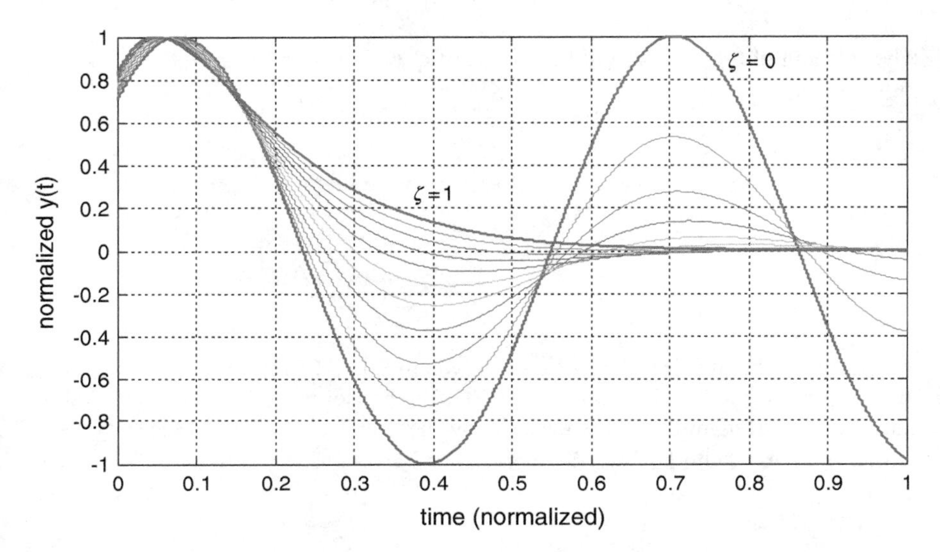

Fig. 7.11 Free response of a second-order system with variation of ζ

Table 7.1 General solution with different damping characteristics

Type of system	Condition	General solution/alternative form
Undamped	$\sigma = 0$	$y(t) = c_1 \cdot e^{j\omega \cdot t} + c_2 \cdot e^{-j\omega \cdot t}$
	$\zeta = 0$	$y(t) = K \cdot \cos(\omega t - \delta)$
Overdamped	Distinct real root $\zeta > 1$	$y(t) = c_1 \cdot e^{\lambda_1 \cdot t} + c_2 \cdot e^{\lambda_2 \cdot t}$
Critically damped	Real double root $\zeta = 1$	$y(t) = (c_1 \cdot + c_2 \cdot t) \cdot e^{\lambda \cdot t}$
Underdamped	$(0 < \zeta < 1)$	$y(t) = e^{\sigma \cdot t}\{k_1 \cos(\omega t) + k_2 \sin(\omega t)\}$
		$y(t) = K \cdot e^{\sigma \cdot t} \cdot \cos(\omega t - \delta)$

Table 7.2 Responses of a second-order system with $\sigma \neq 0$

Distinct real root $a^2 > 4b$	$y(t) = c_1 \cdot e^{\lambda_1 \cdot t} + c_2 \cdot e^{\lambda_2 \cdot t}$
Real double root $a^2 = 4b$	$y(t) = (c_1 \cdot + c_2 \cdot t) \cdot e^{\lambda \cdot t}$
Conjugate pair $a^2 < 4b$	$y(t) = e^{\sigma \cdot t}\{k_1 \cos(\omega t) + k_2 \sin(\omega t)\}$
	$y(t) = K \cdot e^{\sigma \cdot t} \cdot \cos(\omega t - \delta)$

7.7 Internal Stability of the System

The discussion of damping down or amplification of the system leads to describing the system as a stable or unstable system. A system that has damping characteristic in response is a stable system; a system in which response is amplified with time is an unstable system. If we observe the poles of characteristic equation or eigenvalues of state matrix A we can relate the stability of the system due to configuration of the poles. The roots of a second-order polynomial as complex conjugate pair are $\lambda_{1,2} = \sigma \pm j\omega$. Any nth order system formulates in second-order sections because if there is a complex pole then it must exist in conjugate pair. Each pole exhibits its response independently depending upon its characteristic given in Table 7.1 and Table 7.2. In other words it is dependent upon its location in s-plane, which is a complex plane with real parts of poles on the x-axis and imaginary parts on the y-axis. We divide s-plane into three sub-planes to describe the stability of the system. A right half s-plane is termed as unstable plane, a left half s-plane is termed as stable plane and y-axis or $j\omega$ line itself is a marginally stable plane.

Marginally Stable Systems: For undamped or unamplified systems the real parts of a conjugate pair is zero and so the response is neither damped down nor amplified. Both roots lie on y-axis ($j\omega$ line), where one is above the horizontal line and the other is below the horizontal line due to imaginary parts. The sign of imaginary parts changes the initial phase of the output response. The imaginary axis is main frequency axis and so higher the value of ω and so as imaginary part leads to more oscillating behavior of the system. All poles with $\sigma = 0$ are marginally unstable poles and if a single pole exists with $\sigma = 0$, it makes the whole system marginally unstable.

Unstable Systems: A pole of a system exhibits an unstable behavior if the real part is positive or the pole lies in the right half of the s-plane. In a second-order polynomial $D_2(s) = s^2 + a \cdot s + b$, a negative value of a makes a positive real part of pole. An nth order system with any pole with a positive real part makes the whole system an unstable system. The response equations $y(t)$ show that magnitude of $y(t)$ will grow exponentially as $t \to \infty$ and this makes the whole system unstable by adding into net response.

Stable Systems: A system is said to be stable if all the poles of the system lie in the left half of the s-plane. A pole is a stable pole if its real part is negative, which makes the response damped down as $t \to \infty$ eventually. If all poles are on the left side of the s-plane or have negative real parts $\sigma < 0$ then these systems are also called asymptotically stable systems (Fig. 7.12).

A map of s-plane is shown in Fig. 7.13 describing the response profile variation in s-plane. Single pole at origin $s = 0$ gives a constant output response and poles on real axis produce responses without oscillating frequencies. The responses on the right side of origin on real axis grow out towards instability as $t \to \infty$ after starting from non-zero initial conditions. The responses on the left side of origin settle down towards zero as $t \to \infty$ after an excitation by non-zero initial conditions. The right and left side of origin marks the growing out or settling down behavior of the

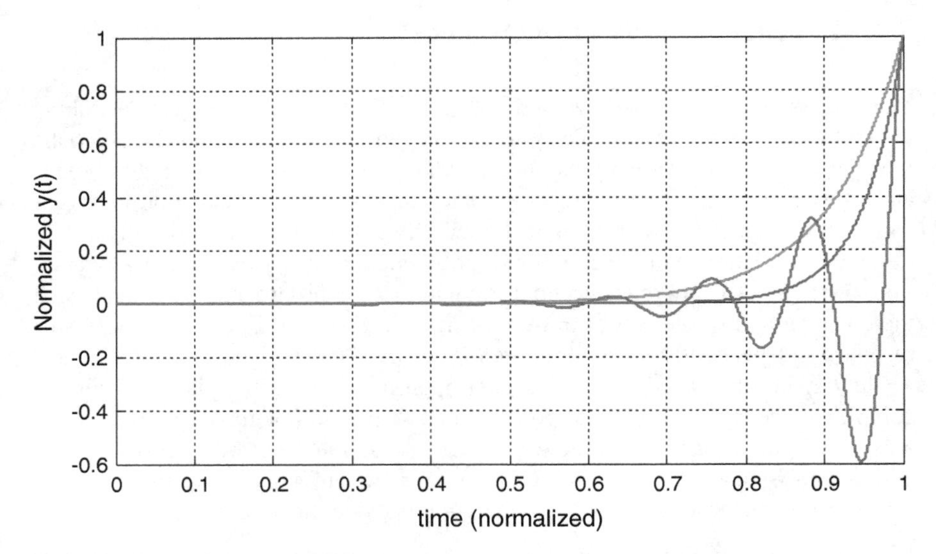

Fig. 7.12 Responses of amplified systems

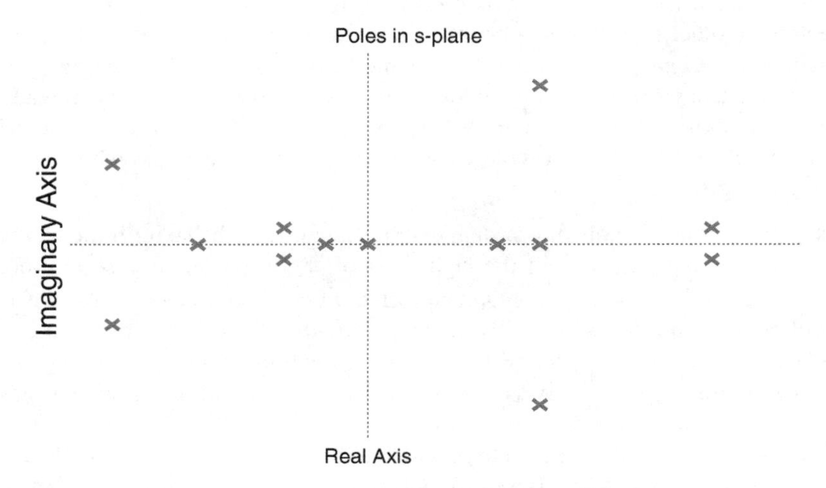

Fig. 7.13 Location of poles in s-plane; poles exist in conjugate pairs except on real axis.

response depicting the stable and unstable nature of the system. The position of imaginary axis shows the oscillating frequency of the system and its amplitude either grows out if it is on the right side or damped down if it is on the left side of the s-plane. As we know, complex roots exist in pairs and so both roots have the same response starting from the same initial conditions on both sides of the real axis. Same responses can be seen at same distant locations of imaginary axis in upper and lower sides of real axis. The upper and lower sides of real axis have nothing to do with the stability of the system, they just describe the amount of oscillatory nature present in the system. If a complex root $z = a + jb$ has $|a| \gg |b|$ then there is quicker

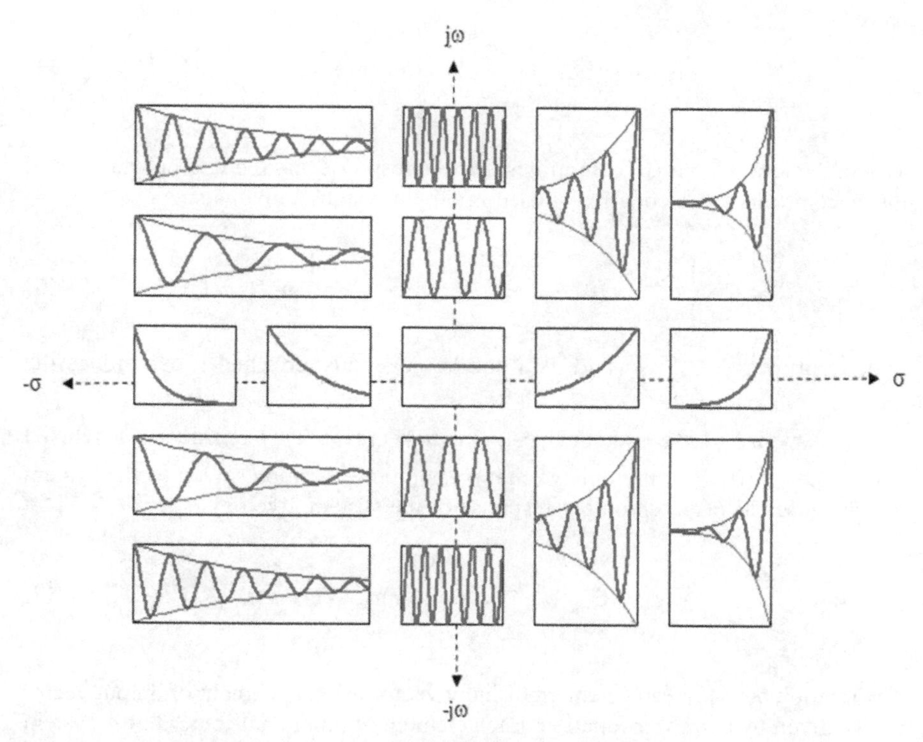

Fig. 7.14 Variation in free responses due to location of a pole in s-plane

damping (or amplification) and lesser oscillations observed over a period of time. On the other hand if $b \gg |a|$ then oscillations are more visible in the response and envelope of response on both sides of s-plane show the speed of convergence or divergence of the system. A conjugate pair of poles on imaginary axis shows an undamped behavior with a constant amplitude sinusoidal behavior at different values of ω. In simple words, internal stability of the system depends upon the location of the pole in the right or left half of the plane. A mirror image of the pole in the left or right half plane stabilizes or destabilizes the system as given in response in Fig. 7.14. A pole in the left half of plane makes the system oscillate with a frequency ω and dampened down towards zero; on the other hand a pole of same magnitude with an opposite sign of real part makes the system oscillate at same frequency and grows the response out of bound.

7.8 Forced Response

In Eq. (7.7) we represented the response of the system in homogenous terms and non-homogenous terms

$$\vec{x}(t) = e^{A \cdot t} \cdot \vec{x}(0) + e^{A \cdot t} \cdot \int_0^t e^{-A \cdot \tau} \cdot B \cdot \vec{u}(\tau) d\tau \tag{7.45}$$

Where the term $e^{A \cdot t} \cdot \vec{x}(0)$ determines the free response and the internal stability of the system. The same equation is used in output equation given as

$$\vec{y}(t) = C \cdot \left\{ e^{A \cdot t} \cdot \vec{x}(0) + \int_0^t e^{A \cdot (t-\tau)} \cdot B \cdot \vec{u}(\tau) d\tau \right\} + D \cdot \vec{u}(t). \tag{7.46}$$

The term $\int_0^t e^{A \cdot (t-\tau)} \cdot B \cdot \vec{u}(\tau) d\tau$ is a non-homogenous term and it determines the forced response of the system due to input vector $\vec{u}(t)$. Considering a relaxed system i.e., $\vec{x}(0) = 0$, the only excitation is due to the input given to the system which causes its effect in output response of the system given by

$$\vec{y}(t) = C \cdot \int_0^t e^{A \cdot (t-\tau)} \cdot B \cdot \vec{u}(\tau) d\tau + D \cdot \vec{u}(t) \tag{7.47}$$

The relation between each element of input vector to each element of output vector is also given by a transfer function. Each element of input contributes to the system and the response then varies accordingly. In order to understand the type of response excited by different inputs, we now consider a stable single input–single output system. If we compare free response term with forced response term, the parameter $e^{A \cdot t}$ as solution of differential equation is common in both parts. We have understood that the stability of the systems depend upon the eigenvalues of the A matrix and so if the system is unstable internally then the force response will also be unstable (if it is not being controlled specifically). There are three responses which are often studied:

- Impulse response
- Step response
- Sinusoidal response or decaying sinusoidal response

We now simulate Fig. 7.4 but instead of taking $u(t) = 0$, we now take $\vec{x}(0) = 0$ in integrator.

7.8.1 Impulse Response

An impulse is a sudden input for a small duration, ideally it is input of infinite magnitude at 0-s but it is not possible to implement it. Dirac's Delta function $\delta(t - a)$ is used as an approximate impulse input which exists in a different application of physical systems. A short duration surge in electrical system, hammer blow, hard landing, kicking or hitting a ball are a few applications which are studied by impulse response. The function $\delta(t-a)$ is given as

$$\delta(t-a) = \lim_{k \to 0} f_k(t-a) \tag{7.48}$$

where

$$f_k(t-a) = \begin{Bmatrix} 1/_k & a \leq t \leq a+k \\ 0 & \text{otherwise} \end{Bmatrix} \tag{7.49}$$

The integral of $f_k(t-a)$ over time is 1 which corresponds to unit impulse response.

$$\int_0^\infty \delta(t-a) = 1 \tag{7.50}$$

We can also note that Laplace transform of impulse function or Dirac's Delta Function is given as

$$\mathcal{L}\{\delta(t-a)\} = \mathrm{e}^{-as} \tag{7.51}$$

For $a = 0$, an impulse function at $t = 0$ has $\mathcal{L}\{\delta(t)\} = 1$ which is used for computation of impulse response of the system. By taking $u(t) = \delta(t)$ and finding an output response is called the impulse response of the system. Note that

$$u(t) = \delta(t) \to U(s) = 1 \tag{7.52}$$

$$\frac{Y(s)}{U(s)} = G(s) \to Y(s) = G(s) \tag{7.53}$$

So the impulse response brings the response of internal dynamics of the system at the output. The Dirac's Delta function has digital equivalent known as Kronecker delta in digital control systems.

Example 7.5 Simulating Impulse Response:

Consider the Examples 7.1–7.4 and plot impulse responses for relaxed systems.

Solution A matrix is different in all examples for undamped (UND), overdamped (OD), critically damped (CD), and underdamped (UD) systems. We simulate the system using the Simulink diagram on Fig. 7.15 for relaxed systems.

$$A_{\mathrm{UND}} = \begin{bmatrix} 0 & -1 \\ 1 & 0 \end{bmatrix} \quad A_{\mathrm{OD}} = \begin{bmatrix} -1 & 0 \\ 0 & -2 \end{bmatrix}$$

$$A_{\mathrm{CD}} = \begin{bmatrix} -1 & 1 \\ 0 & -1 \end{bmatrix} \quad A_{\mathrm{UD}} = \begin{bmatrix} -1 & 1 \\ -1 & -1 \end{bmatrix}$$

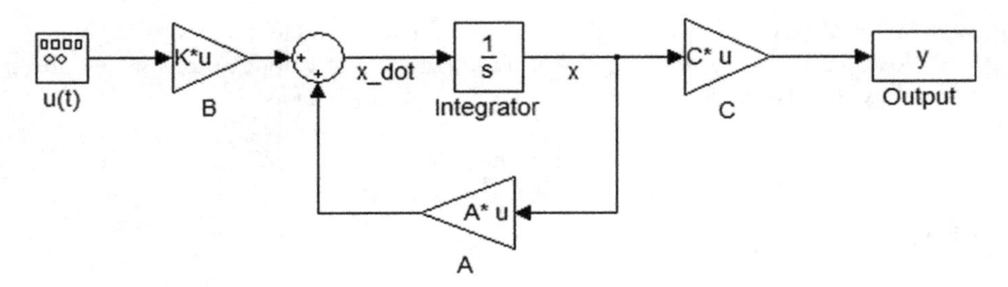

Fig. 7.15 Simulation diagram for force response

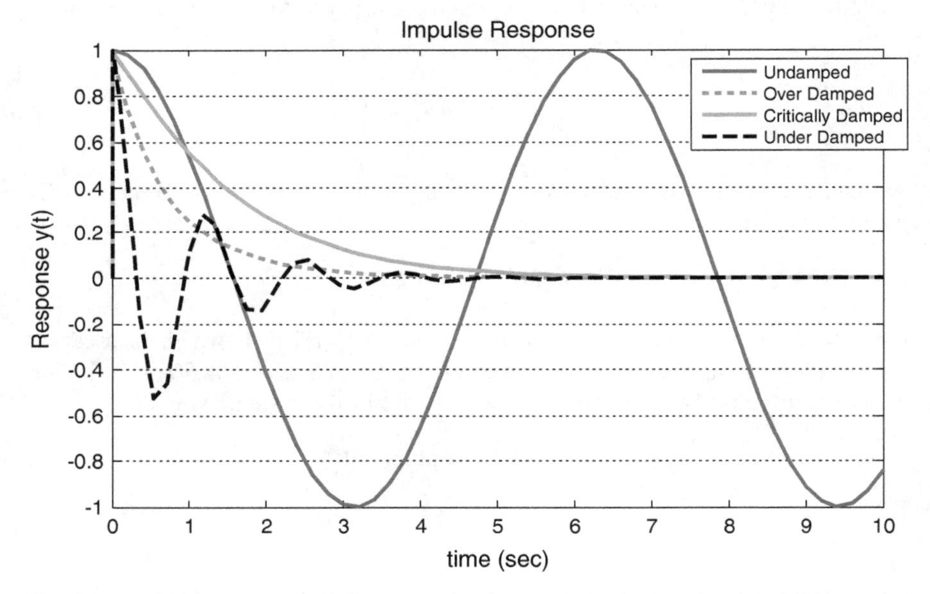

Fig. 7.16 Impulse response of Undamped, Overdamped, Critically Damped and Under Damped System

Impulse response is created by using "Signal Builder," block for a Dirac's delta function (Fig. 7.16).

We note that impulse response shows the internal dynamics of the system as shown in Fig. 7.17. The impulse response of unstable systems will also show the amplification in the magnitudes growing out of bound. A system hit by a huge impulse force excites the internal dynamics of systems which later settle down just like excitation with initial conditions.

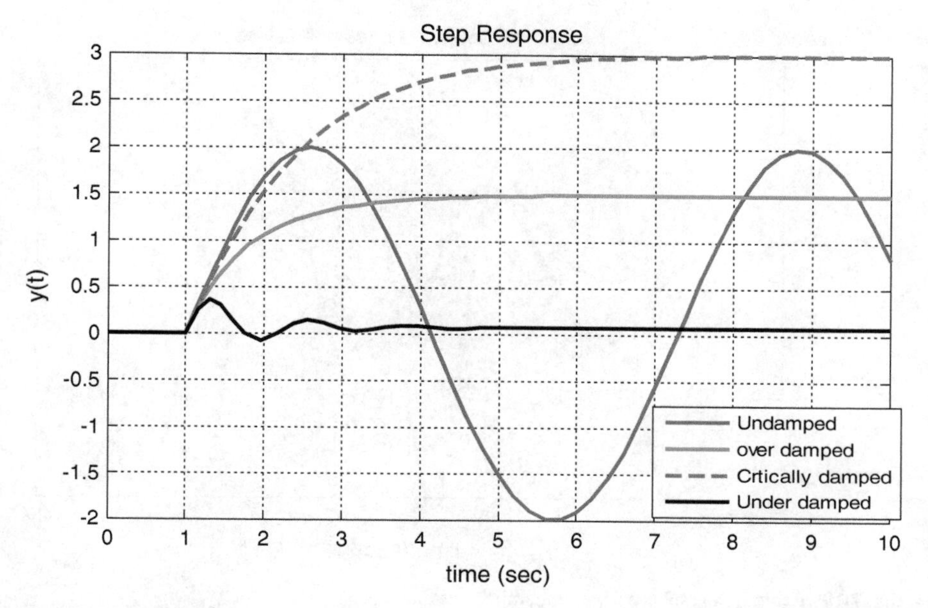

Fig. 7.17 Step response of an undamped, overdamped, critically damped, and underdamped system

7.8.2 Step Response

A step response of the system is evaluated after giving a constant step input usually starting at $t = 0$. A step function is given as

$$u(t - t_i) = \begin{Bmatrix} 1 & t \geq t_i \\ 0 & \text{otherwise} \end{Bmatrix} \tag{7.54}$$

The Laplace transform of a step function given in (7.53) is

$$\mathcal{L}\{u(t - a)\} = \frac{1}{s}\mathrm{e}^{-t_i s} \tag{7.55}$$

The transfer function of the system with a step input $(t - a)$ can be represented as

$$Y(s) = G(s) \cdot U(s) = \frac{G(s)}{s}\mathrm{e}^{-t_i s} \tag{7.56}$$

The output $y(t)$ is like a free response of a system with an additional pole at $s = 0$ starting with a delay of t_i seconds. The parameter $\mathrm{e}^{-t_i s}$ in the transfer function represents the delay in the system. The application of a step response analysis arises

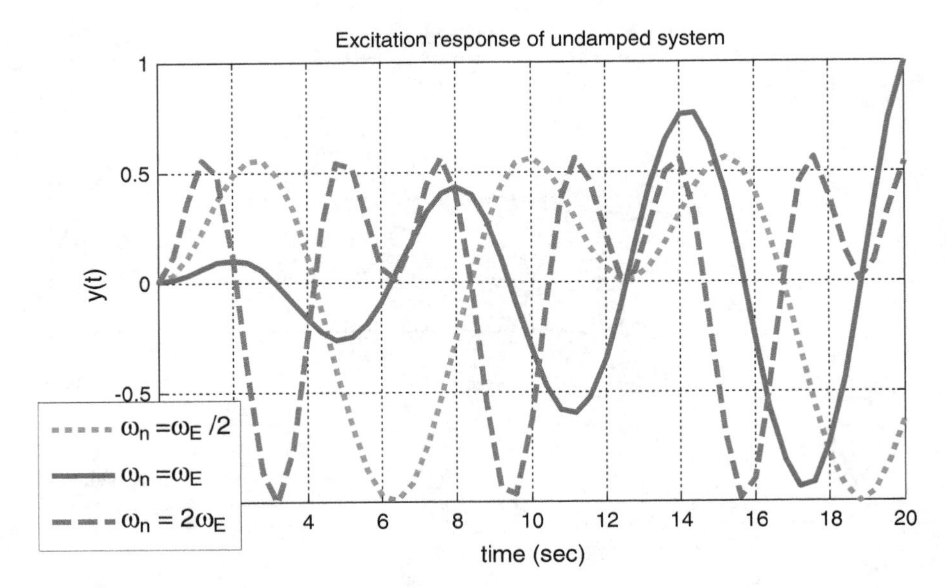

Fig. 7.18 Excitation response of an undamped system

when a system is hit by a constant force over a period of time of operation like combustion, constant power supply, constant force exerting on a spring, etc.

Example 7.6 Simulating Step Response

Find a step response of the all four systems in previous Example 7.5 with step input $u(t-1)$ and discuss responses.

Solution A step function of MATLAB Simulink can be used in Fig. 7.15 to determine the step responses of the undamped, overdamped, critically damped, and underdamped systems. Figure 7.18 shows the responses of the systems.

An undamped system oscillates with twice the magnitude as three poles now lie on y-axis starting at $t_i = 1$ seconds. The remaining three systems settle at a constant value but not at zero steady-state. The constant input makes the system constant at non-zero position depending upon the location of poles in s-plane. An underdamped system may get lower magnitude while settling close to zero when the real part is smaller, on the other hand if real part is larger than it may also settle at a higher constant value.

7.8.3 Sinusoidal Response

The dynamic response of the system with a forcing sinusoidal function as an input has a variety of parameters to be discussed. An input with a frequency component excites the system and the system has a combination of frequencies in its output response. A forced input with magnitude F and excitation frequency ω_E can be given as

$$u(t) = F \cdot \cos(\omega_E t) \tag{7.57}$$

The Laplace transform of the function is

$$U(s) = \frac{F}{s^2 + \omega_E^2} \tag{7.58}$$

The output response in a transfer function is now given as

$$Y(s) = G(s) \cdot \frac{F}{s^2 + \omega_E^2} \tag{7.59}$$

We can observe that the system will have two additional poles on imaginary axis at $\pm j\omega_E$ and the combination of these poles with poles of $G(s)$ affects the response of the overall system. A transfer function $G(s)$ has natural frequency ω_n and the location of poles due to ω_n and $\pm j\omega_E$ defines the response of the system. A second-order system may be easily studied in order to check the different responses.

7.9 Resonance

When excitation frequency approaches the natural frequency i.e., $\omega_E \rightarrow \omega_n$ then the magnitude of the system approaches infinity. An undamped system operating at $\omega_E = \omega_n$ becomes an unstable system because of double pole at the same location on the imaginary axis. Remember these poles are in conjugate pair, a double pole on imaginary axis in upper half plane {above x-axis} and a double pole on imaginary axis in lower half plane {below x-axis} of s-plane. A double pole on the imaginary axis is an unstable system, so the undamped system becomes an unstable system when excited at the natural frequency. But when an undamped system is excited by sinusoidal input with a different frequency response $y(t)$ contains both frequencies: natural frequency ω_n as well as excitation frequency ω_E. The magnitude of the response also varies due to variation in ω_E (Fig. 7.19).

The responses of overdamped and critically damped systems are similar under excitation from sinusoidal frequency because the poles of both systems do not have imaginary parts. The system oscillates at the excitation frequency ω_E after an initial overshoot in magnitude. The normalized response plot of overdamped and critical damped system shown in Fig. 7.20 depicts the initially amplified cycle and then almost constant amplitude for both systems in both excitation frequencies ω_E and 2 ω_E. There is no concept of resonance in overdamped and critically damped systems because of damping and no dominating natural frequency component in the poles. The response of an underdamped system creates a scenario of practical resonance because the real systems have some damping in them. If a real part of

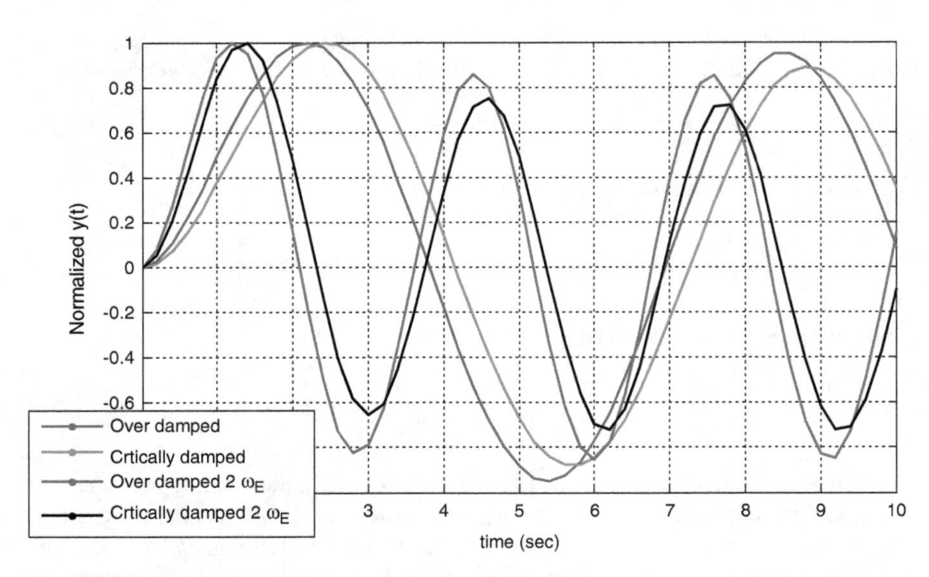

Fig. 7.19 Excitation response of overdamped and critically damped systems with frequency ω_E

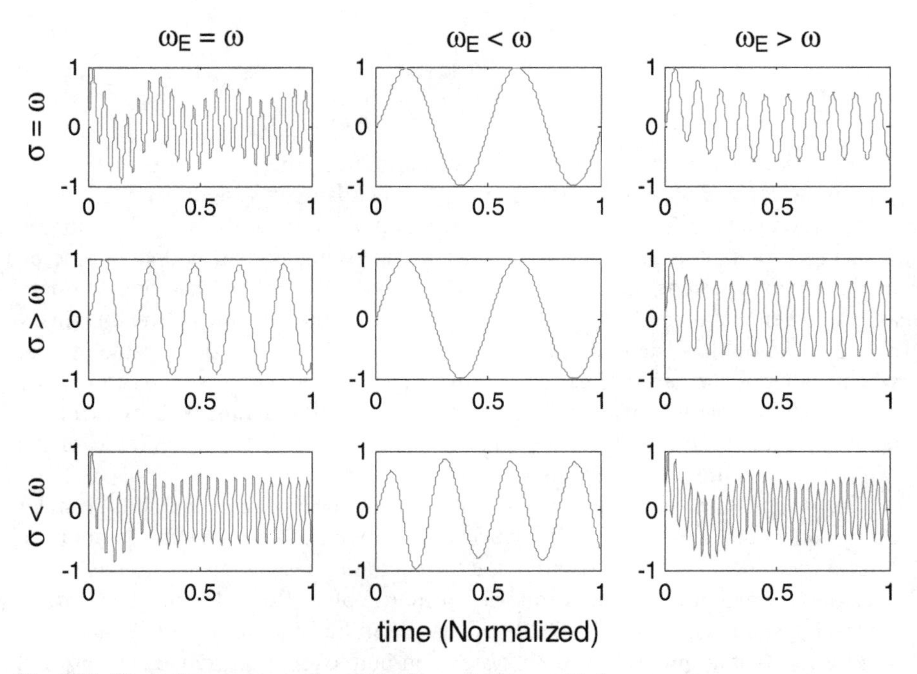

Fig. 7.20 Excitation response of an underdamped system with frequency ω_E

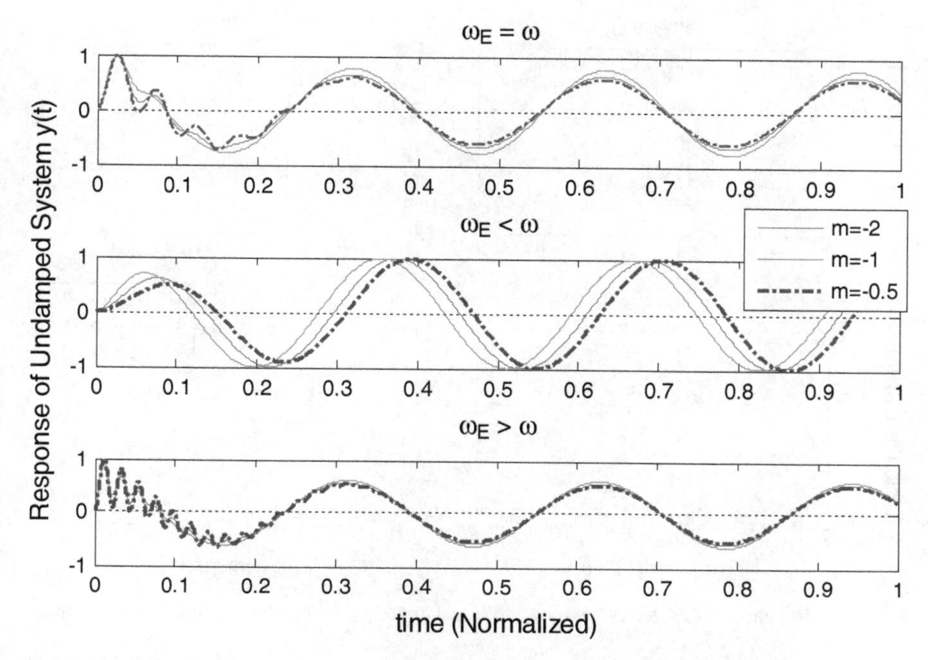

Fig. 7.21 Excitation response of an undamped system with decaying sinusoidal input

eigenvalues of A matrix is very small than imaginary part, it means there is less damping in the system. This system may also resonate or oscillate with high amplitude when excited by frequency equal to the imaginary part {natural frequency} of the system. Due to damping effect, the system settles at oscillating frequency after following a transient response in the first few cycles (Fig. 7.21).

7.10 Decaying Sinusoidal Response

A decaying sinusoidal input signal is represented as an exponentially decaying oscillating function exactly like the underdamped response of a system. In practical systems, a controlled input is also decreasing and settling at a constant or zero value by attaining desired objectives. An input function is given as

$$u(t) = \mathrm{e}^{mt} \cos(\omega_E t) \tag{7.60}$$

Where $m < 0$ is the decaying ratio, and if $m = 0$, the input signal becomes pure sinusoidal and for $m > 0$ the input signal will be amplified with time; ω_E is the excitation frequency. Figure 7.22 shows that the undamped system settles at oscillating frequency after transient response and the higher the magnitude of $-m$, the quicker the settling time (Fig. 7.23).

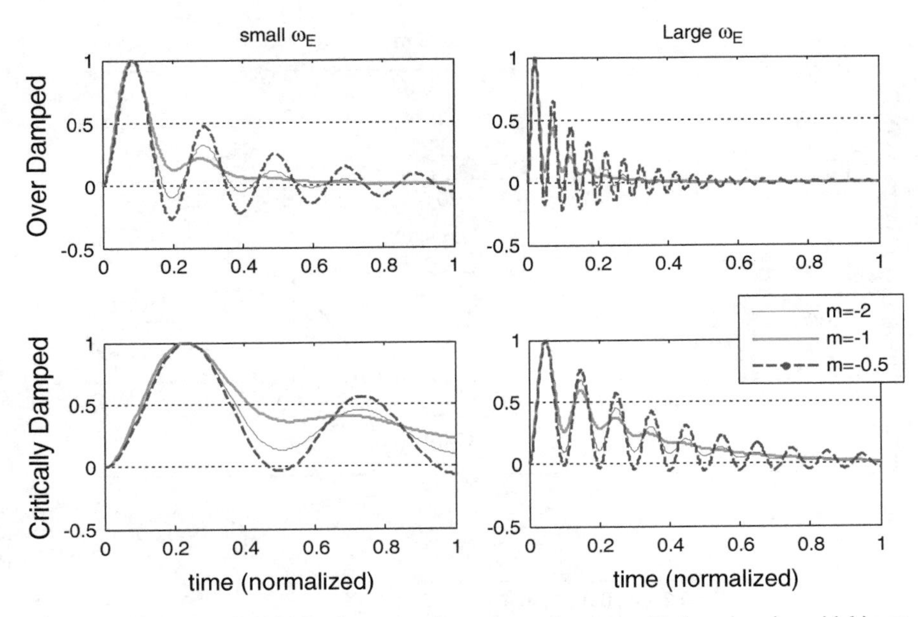

Fig. 7.22 Excitation of critically damped and overdamped system with decaying sinusoidal input

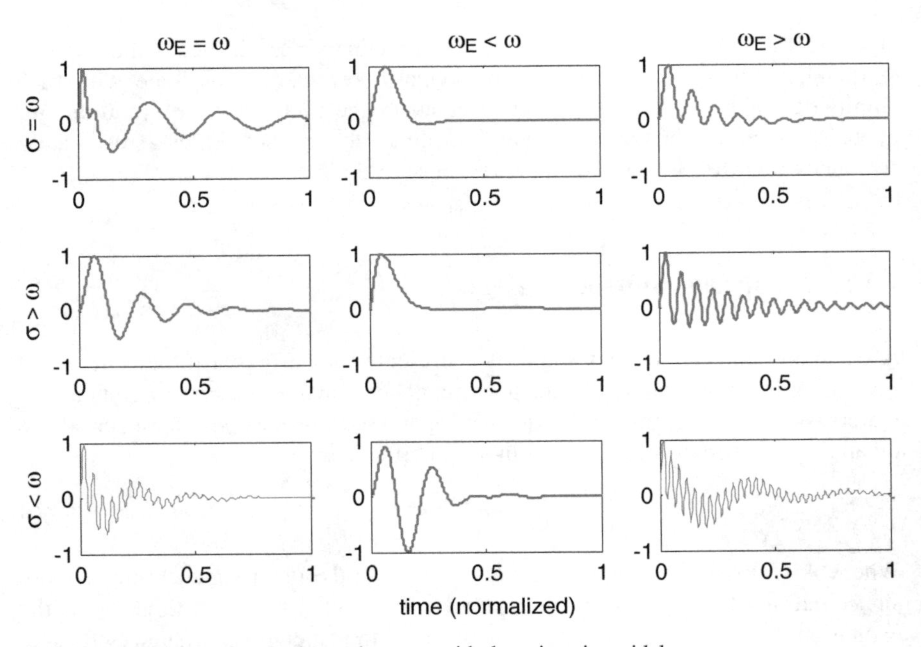

Fig. 7.23 Excitation of underdamped system with decaying sinusoidal response

Overdamped and underdamped systems behave similarly as both settle towards zero in steady-state response after oscillatory transient initially. An overdamped system settles earlier than the critically damped systems if poles of both systems are closer in left half s-plane. The large value of excitation frequency makes the system oscillate more and steady-state also have frequent peaks whereas small excitation frequency oscillates the system with smaller frequency but larger amplitude approximately at same instances of time. Underdamped systems are more realistic systems which has damping and settling at steady-state due to both input and internal dynamics. For a system with a real part of pole equal to imaginary part and excitation frequency equal to imaginary part; response will take longer to settle. If real part (damping) is greater and excitation frequency is smaller than imaginary part of the pole, the system will take less time to settle. We can also note that oscillation in the response occurs either due to larger excitation frequency or smaller damping coefficient of the system. Damping ratio in poles and input tries to stabilize the system earlier whereas excitation frequency and imaginary part of the poles oscillate the system and make it harder to settle earlier (Fig. 7.24).

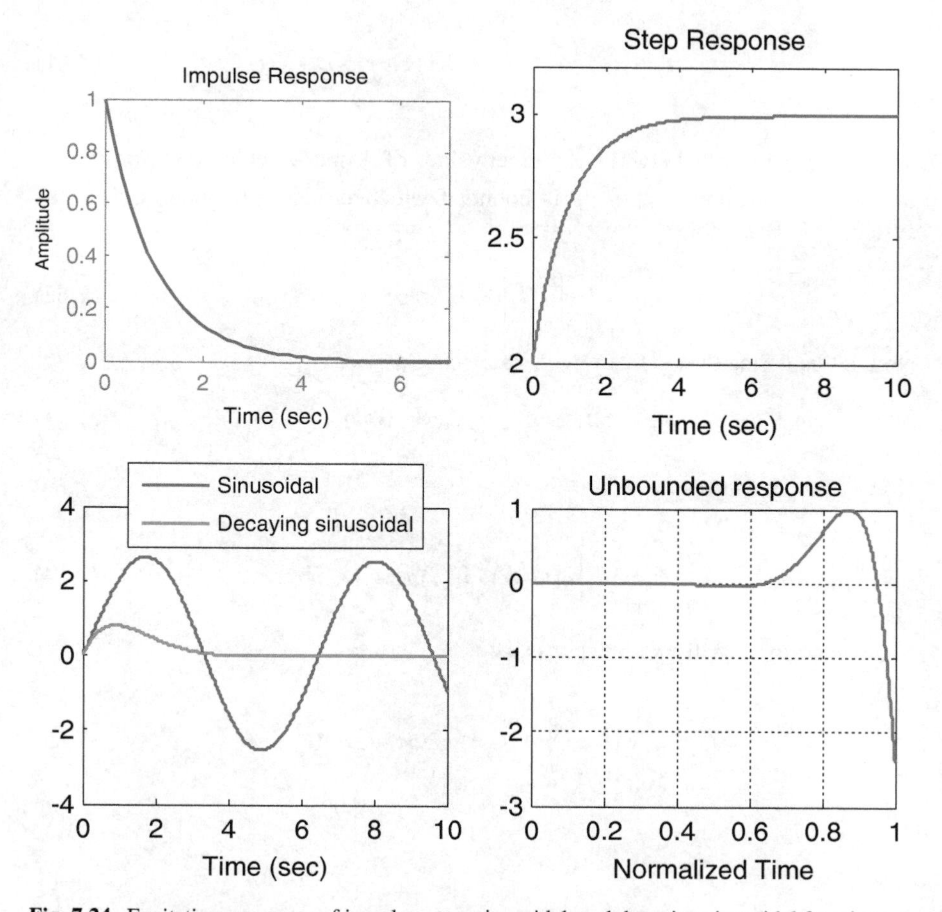

Fig. 7.24 Excitation responses of impulse, step, sinusoidal, and decaying sinusoidal functions as bounded input and a sinusoidal function as unbounded input for system in Eq. (7.71)

There are different variants of responses due to the combination of slower and faster eigenvalues of the system. Eigenvalues farther from origin on the x-axis are faster {large value of damping or amplification} and those closer to origin are slower to settle. It should also be noted that the third- and higher order systems usually have a combination of different poles and thus the response of linear systems also adds up accordingly.

7.11 External Stability

A system is internally stable if the eigenvalues are in the left half plane, but a stability related with input of the system is known as external stability. This stability is related with forced response and also known as Bound Input Bounded Output {BIBO} stability of the system. This stability determines the response of the system with bounded input. The forced response of the system is given as

$$\vec{y}(t) = C \cdot \int_0^t e^{A\cdot(t-\tau)} \cdot B \cdot \vec{u}(\tau)d\tau + D \cdot \vec{u}(t) \tag{7.61}$$

We determine internal stability by eigenvalues of A matrix which also contributes in the forced response. If $\vec{u}(t)$ is bounded which produces bounded output the system is BIBO stable

$$\left|\vec{u}(t)\right| < U < \infty \tag{7.62}$$

Now let us define the realization of state space as

$$g(t-\tau) = C \cdot e^{A\cdot(t-\tau)} \cdot B \tag{7.63}$$

Now we represent response as

$$\vec{y}(t) = \int_0^t g(t-\tau) \cdot \vec{u}(\tau)d\tau + D \cdot \vec{u}(t) \tag{7.64}$$

A bounded input will yield response as

$$\vec{y}(t) = \int_0^t g(t-\tau) \cdot U d\tau + D \cdot U \tag{7.65}$$

As $D < \infty$ is a real bounded matrix so the response $\vec{y}(t)$ will be bounded if

$$\left|\vec{y}(t)\right| = \left|\int_0^t g(t-\tau)\cdot\vec{u}(\tau)d\tau + D\cdot\vec{u}(t)\right| < \infty. \quad (7.66)$$

As we know that both $\vec{u}(\tau)$ and D are bounded so

$$\left|\vec{y}(t)\right| = \left|\int_0^t g(t-\tau)\cdot\vec{u}(\tau)d\tau + D\cdot\vec{u}(t)\right| \quad (7.67)$$

$$\left|\vec{y}(t)\right| = \left|\int_0^t g(t-\tau)\cdot U d\tau + D\cdot U\right| \quad (7.68)$$

$$\left|\vec{y}(t)\right| = |U|\cdot\left|\int_0^t g(t-\tau)\cdot d\tau\right| + |D\cdot U| \quad (7.69)$$

So $\left|\vec{y}(t)\right|$ will be bounded if and only if $\left|\int_0^t g(t-\tau)\cdot d\tau\right|$ is bounded

$$\left|\vec{y}(t)\right| < \infty \text{ iff } \left|\int_0^t g(t-\tau)\cdot d\tau\right| < \infty. \quad (7.70)$$

An external stability also depends upon realization of transfer function or state space formulation of problem, or in other words related with internal stability of the system. A question arises that if the input is bounded then how is external stability different than internal stability? Internal stability depends upon eigenvalues of the system and external stability also depends upon the same $G(s)$ or $g(t)$, and if the system is internally unstable then it will be externally unstable as well. This is true in most cases but a difference occurs due to the realization of the system, there may be some zeros in the system that are cancelling the unstable poles and thus making the system stable for specific input conditions. Alternatively, in the state space realization the unstable eigenvalues may not be contributing in the output response of the system. All inputs studied in the previous section are bounded inputs and so forced response of the system is determined as BIBO response and displaying the external stability of the system.

Example 7.7 Simulating Sinusoidal Response

Consider the following system and determine its internal and external stabilities. Plot the different responses of the system with bounded and unbounded inputs.

$$\dot{\vec{x}}(t) = \begin{bmatrix} -1 & 0 \\ 2 & 1 \end{bmatrix} \cdot \vec{x}(t) + \begin{bmatrix} 1 \\ 1 \end{bmatrix} \cdot u(t) \tag{7.71}$$
$$y(t) = \begin{bmatrix} 1 & 0 \end{bmatrix} \cdot \vec{x}(t) + [2] \cdot u(t)$$

Solution Internal stability of the system is determined by the eigenvalues of the state matrix A. The eigenvalues of the A matrix are 1 and -1 {using eig(A)} which clearly show that one of the eigenvalues is in right half plane and so the system is internally unstable. In order to determine external stability we need to look the transfer function i.e.,

$$G(s) = \frac{2s^2 + s - 3}{s^2 - 1} = \frac{(s-1)(s+3/2)}{(s+1)(s-1)} = \frac{(s+3/2)}{(s+1)} \tag{7.72}$$

We can see that unstable pole has been canceled with a zero and the remaining system is actually a stable first-order system. If we observe C matrix which shows that only first state of the system is contributing in the output and first state of the system is an independent stable eigenvalue of the system {decomposition of states}. This system is internally unstable but externally stable when a bounded input is given to the system. We now plot the bounded responses of the system with respect to four bounded inputs with an unbounded output. Figure 7.25 shows the corresponding responses computed from the following MATLAB commands

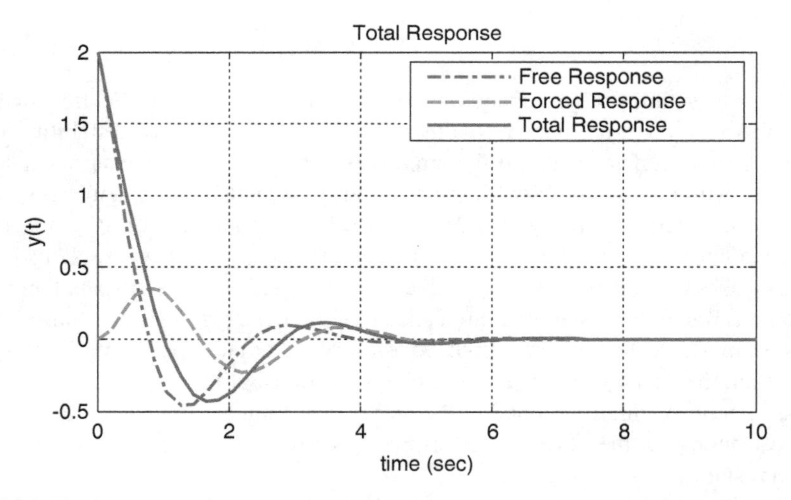

Fig. 7.25 Total response of a system as combination of free response and forced response

```
A=[-1 0;1 1];
B=[1;1];
C=[1 0];
D=2;
sys=ss(A,B,C,D);
t=0:0.01:10;
u_impulse=[1; zeros(length(t)-1,1)];
u_step=ones(1,length(t));
u_sin=sin(1*t);
u_dsin=exp(-t).*sin(1*t);
u_ub=exp(t).*sin(1*t);
y_impulse=lsim(sys,u_impulse,t);
y_step=lsim(sys,u_step,t);
y_sin=lsim(sys,u_sin,t);
y_dsin=lsim(sys,u_dsin,t);
y_ub=lsim(sys,u_ub,t);
subplot(2,2,1);
impulse(sys);
subplot(2,2,2);
plot(t,y_step,'LineWidth',2);
xlabel('Time (sec)');
title('Step Response');
subplot(2,2,3);
plot(t,y_sin,'LineWidth',2); hold on
plot(t,y_dsin,'r','LineWidth',2);
title('Sinusoidal Response');
legend('Sinusoidal','Decaying sinusoidal');
xlabel('Time (sec)');
subplot(2,2,4)
plot(t/max(t),y_ub/max(y_ub),'LineWidth',2);
grid on
xlabel('Normalized Time');
title ('Unbounded response');
```

7.12 Total Response of the System

A complete response of the system is comprised of free response as well as forced response of the system as already given by the equation

$$\vec{y}(t) = C \cdot \left\{ e^{A \cdot t} \cdot \vec{x}(0) + \int_0^t e^{A \cdot (t-\tau)} \cdot B \cdot \vec{u}(\tau) d\tau \right\} + D \cdot \vec{u}(t) \tag{7.73}$$

If a system is not relaxed then $\vec{x}(0) \neq 0$ and given initial condition excites the free response of the system, a non-zero input vector $\vec{u}(t)$ excites the forced response of the system. Total response is the linear combination of both responses, a superposition of homogenous and particular solution of differential equation. The final stability is the combination of internal and external stability of the system and profile of response also depends upon combination of both. We studied four types of free responses of second-order system, and four types of bounded input forced responses, which show that different combinations of responses are possible depending upon the location of eigenvalues as well as the type of input given. The same eigenvalues of the system or poles of characteristic contribute to free and forced response so usually there are similar types of responses that combine to produce an overall effect. Some examples are the combination of a resonant forced response and undamped system, the combination of an overdamped system with decaying sinusoidal input, etc.

Example 7.8 Simulating Total Response

Simulate and plot the free response, forced response, and total response of the system

$$A = \begin{bmatrix} -1 & 2 \\ -2 & -1 \end{bmatrix}, \ B = \begin{bmatrix} 1 \\ 1 \end{bmatrix}, \ C = [1 \quad 1], \ D = [1] \tag{7.74}$$

$$\vec{x}(0) = \begin{bmatrix} 1 \\ 1 \end{bmatrix} \tag{7.75}$$

$$u(t) = \mathrm{e}^{-t} \sin(2t) \tag{7.76}$$

Solution The total response of the system is a combination of free response and forced response. Simulating Fig. 7.1 with plant dynamics given in Eq. (7.73), initial conditions given in Eq. (7.74) and decaying sinusoidal input given in Eq. (7.75), free response and forced response given in Fig. 7.4 and Fig. 7.15 respectively yields the total response by giving non-zero initial conditions in Fig. 7.15 (modified diagram of Fig. 7.1). We can observe that free response from magnitude of 2 for $y(t)$ and zero magnitude in forced response. The total response also starts with magnitude 2 and settles earlier due to out of phase free and forced responses (Fig. 7.26).

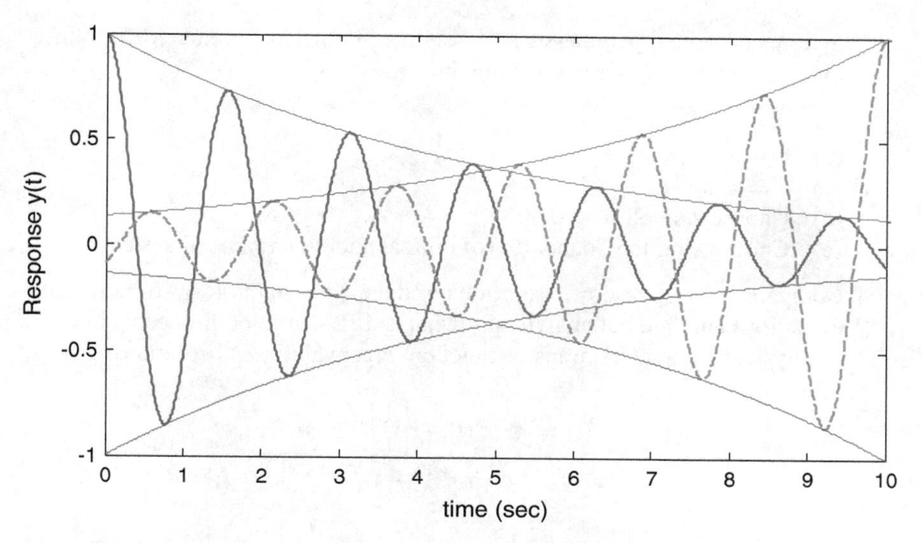

Fig. 7.26 Comparison of responses for the eigenvalue of same magnitude and frequency in right and left half of the s-planes

Problems

P7.1 Complex eigenvalues or complex poles always appear in pairs but do not appear in matrix A or coefficients of characteristic equation $D(s) = 0$. Analyze third- and fourth-order polynomial equations and show that possibilities of different complex roots {poles} in both equations. How you relate this to state matrix A?

P7.2 A harmonic oscillator, Equation of a mass spring system without a damper are given as $m\ddot{y}(t) + ky(t) = 0$. Find a solution of this oscillator using Laplace transform and find its natural frequency. Represent this system in state space formulation and simulate it using the following parameters: 100 N force pulls the system for 0.995 m in relaxed position. What will be the response of the system for position and velocity if we pull the system to following situations and release? {Use $g = 9.81\,\text{m/s}^2$}

(a) 0.25 m only with negligible speed
(b) 0.25 m with 1 m/s

P7.3 A harmonic oscillator will keep on moving forever if there is no damping in it but all real systems have damping and in the presence of damping the equation of motion is given as

$$m\ddot{y}(t) + c\dot{y}(t) + ky(t) = 0$$

Represent the system in state space formulation, use Laplace transform to solve the system and find out its damping ratio and damped natural frequency in terms of system parameters m, c and k. Simulate and relate the responses to the undamped systems for following cases

(a) $c = 80\,\text{kg/s}^2$
(b) $c = 64\,\text{kg/s}^2$
(c) $c = 40\,\text{kg/s}^2$
(d) Choose a small $c \approx 0.001$
(e) Choose c on the boundaries of critical and underdamped system

P7.4 Compare the response of two critically damped systems for different initial conditions and find out analytically what is different in both systems in terms of physical parameters, transfer function, eigenvalues, and solution

$$\dot{\vec{x}}(t) = A \cdot \vec{x}(t) + \begin{bmatrix} 1 \\ 1 \end{bmatrix} \cdot u(t)$$
$$y(t) = \begin{bmatrix} 1 & 1 \end{bmatrix} \cdot \vec{x}(t)$$

$$\vec{x}_{01}(0) = \begin{bmatrix} 1 \\ 1 \end{bmatrix} \quad \vec{x}_{02}(0) = \begin{bmatrix} -1 \\ -1 \end{bmatrix} \quad \vec{x}_{03}(0) = \begin{bmatrix} -1 \\ 1 \end{bmatrix} \quad \vec{x}_{04}(0) = \begin{bmatrix} 1 \\ -1 \end{bmatrix}$$

$$A = \begin{bmatrix} -1 & 1 \\ 0 & -1 \end{bmatrix} \quad \text{and} \quad A = \begin{bmatrix} -1 & 0 \\ 0 & -1 \end{bmatrix}$$

P7.5 Find the transfer function of the system given in P2, P3, and P4 using MATLAB. Find the natural frequencies, damping ratios of all systems and use MATLAB for Bode plot of all the systems in one figure. Compare damping ratios of the system in bode plot and relate with responses given in Fig. 7.8. What happens if the numerator of the system changes but the denominator remains the same in Bode plot?

P7.6 Find transfer functions and plot the simulated free responses of the following system with different initial conditions. Also plot the envelopes of oscillatory systems and find out how the initial or maximum point of the envelope is calculated.

(a)

$$\dot{\vec{x}}(t) = \begin{bmatrix} 0 & -2 \\ 2 & 0 \end{bmatrix} \cdot \vec{x}(t) + \begin{bmatrix} 1 \\ 2 \end{bmatrix} \cdot u(t)$$
$$y(t) = \begin{bmatrix} 2 & 1 \end{bmatrix} \cdot \vec{x}(t)$$

$$\vec{x}_{01}(0) = \begin{bmatrix} 1 \\ 0 \end{bmatrix} \quad \vec{x}_{02}(0) = \begin{bmatrix} 0 \\ 1 \end{bmatrix} \quad \vec{x}_{03}(0) = \begin{bmatrix} 1 \\ 1 \end{bmatrix}$$

(b)

$$\dot{\vec{x}}(t) = \begin{bmatrix} 2 & 0 \\ 0 & 3 \end{bmatrix} \cdot \vec{x}(t) + \begin{bmatrix} 2 \\ 1 \end{bmatrix} \cdot u(t)$$
$$y(t) = \begin{bmatrix} 1 & 2 \end{bmatrix} \cdot \vec{x}(t)$$

$$\vec{x}_{01}(0) = \begin{bmatrix} 10 \\ 1 \end{bmatrix} \quad \vec{x}_{02}(0) = \begin{bmatrix} -1 \\ -10 \end{bmatrix} \quad \vec{x}_{03}(0) = \begin{bmatrix} -2 \\ 1 \end{bmatrix} \quad \vec{x}_{04}(0) = \begin{bmatrix} 2 \\ -1 \end{bmatrix}$$

(c)

$$\dot{\vec{x}}(t) = \begin{bmatrix} 2 & 1 \\ 0 & 2 \end{bmatrix} \cdot \vec{x}(t) + \begin{bmatrix} 1 \\ 1 \end{bmatrix} \cdot u(t)$$
$$y(t) = \begin{bmatrix} 1 & 1 \end{bmatrix} \cdot \vec{x}(t)$$

$$\vec{x}_{01}(0) = \begin{bmatrix} 1 \\ 1 \end{bmatrix} \quad \vec{x}_{02}(0) = \begin{bmatrix} -1 \\ -1 \end{bmatrix} \quad \vec{x}_{03}(0) = \begin{bmatrix} -1 \\ 1 \end{bmatrix} \quad \vec{x}_{04}(0) = \begin{bmatrix} 1 \\ -1 \end{bmatrix}$$

(d)

$$\dot{\vec{x}}(t) = \begin{bmatrix} 2 & 3 \\ -3 & 2 \end{bmatrix} \cdot \vec{x}(t) + \begin{bmatrix} 1 \\ 1 \end{bmatrix} \cdot u(t)$$
$$y(t) = \begin{bmatrix} 2 & 2 \end{bmatrix} \cdot \vec{x}(t)$$

$$\vec{x}_{01}(0) = \begin{bmatrix} 2 \\ 1 \end{bmatrix} \quad \vec{x}_{02}(0) = \begin{bmatrix} -1 \\ -2 \end{bmatrix} \quad \vec{x}_{03}(0) = \begin{bmatrix} -1 \\ 2 \end{bmatrix} \quad \vec{x}_{04}(0) = \begin{bmatrix} 2 \\ -1 \end{bmatrix}$$

Normalize the time and response amplitude to the maximum values and compare your plots without normalization; why does normalization sometimes produce better understanding? Is there any other parameter which may be used as a normalization factor for response? Find the eigenvalues of all systems and find out whether the systems are stable, unstable, or marginally stable based upon eigenvalues and free responses.

References

1. Kreyszig, Erwin. 2006. *Advanced Engineering Mathematics*, 9th ed. Hoboken: Wiley.
2. Karnopp, Dean C., Donald L. Margolis, and Ronald C. Rosenberg. 2012. *System Dynamics—Modeling and Simulation of Mechatronics Systems*, 5th ed. Hoboken: Wiley.
3. Friedland, Bernard. 2005. *Control System Design—An Introduction to State Space Methods*. New York: Dover.
4. Chen, Chi-Tsong. 1999. *Linear System Theory and Design*, 3rd ed. New York: Oxford University Press.

Chapter 8
Introduction to Control Systems

A control system is usually defined as a device or set of devices in order to manage, direct, regulate, or command the other set of devices. Typically everything in nature is working under some control mechanisms because every single item is controlled by some higher authority for desired objectives. The objectives or goals are the desired outputs of the system, and whatever resources are needed in order to achieve these objectives are inputs to the system. The authority that manages the resources to achieve desired objective for the system is called the "control system." Here we are talking about two systems, one system that is being controlled normally called a "plant," and another system that is controlling called the "controller." Both are defined as a complete system individually because of different inputs and outputs. In reality, the outputs of a control system become the inputs of the plant and the outputs of the plant become the inputs to the control system. So, this loop works the objectives are achieved, or the resources are depleted. The overall system in this loop configuration, as shown in Fig. 8.1, is called the feedback system, closed loop system, or automatic system.

Figure 8.1 shows that a system under control or plant receives inputs or energy from a resource pool and operates to meet the objectives. There are sensors that measure the current state of the system and pass to the control system or control which computes guided ways in order to meet objectives and thus steer the resource pool to provide power (energy, resources) to the plant by its actuating mechanism. This scheme is explained with different types of block diagrams to represent this loop. A system of Fig. 8.1 given as simplified simulation diagram in Fig. 8.2 with actuators and measurement sensors is represented as the transfer function.

Normally frequency responses of actuators are constant over the desired frequency range, and so these transfer functions are taken as unity, i.e., $A(s) = 1$. This leads to the statement that input to actuators is equal to input to plant i.e., $u_{\mathrm{a}}(t) = u_{\mathrm{p}}(t)$. If there is no reference input i.e., $r(t) = 0$ then it further simplifies that input to plant is the direct output of control system $u(t)$. Measurement sensors are also chosen such that their frequency responses are constant i.e., $M(s) = 1$, but most likely our true objectives are not directly measurable. What we can measure

A.M. Mughal, *Real Time Modeling, Simulation and Control of Dynamical Systems*,
DOI 10.1007/978-3-319-33906-1_8

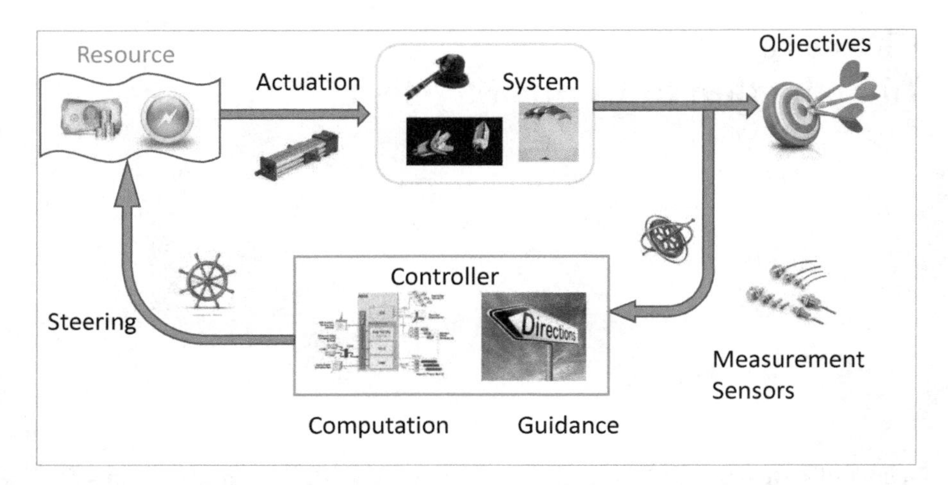

Fig. 8.1 A loop of plant and control systems

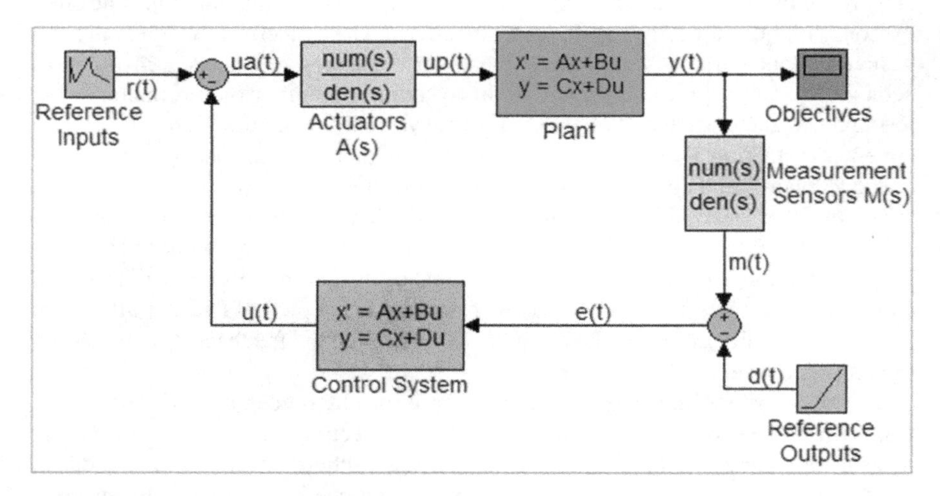

Fig. 8.2 Simulation block diagram of control loop

from the behavior of plant $m(t)$ may be different than what we actually want to achieve as objectives $y(t)$. Reference output is the desired output $d(t)$ at sensors with the error calculated between actual measurements and desired (reference) outputs as error signals $e(t)$, which become input to control systems. We can summarize the input and output of the closed loop system as

Inputs: There are two types of inputs to the closed loop systems.

Controllable inputs $u(t)$: An output of a control system that drives the plant.

Exogenous inputs: These are either reference inputs or disturbances in the system that are due to external sources or internal dynamics, e.g., delays or approximation errors of modeling, etc. These inputs are not controllable by the control system and add the burden to achieve desired objectives in required time.

Reference inputs: These inputs are the desires for the system; a control system tries to achieve these but cannot control these inputs, which is why these are also termed as exogenous inputs. Reference input $r(t)$ and reference output $d(t)$ are both reference inputs to the closed loop feedback system.

Outputs: There are two types of outputs of the plant, one that can be measured through sensors known as measurements $m(t)$, and the other, which are desired objectives of the plant $y(t)$. In a simplified system, measurements are usually taken as desired objectives which yield $y(t) = m(t)$.

8.1 Representation of Controller

A Controller can be represented as a transfer function from error $e(t)$ between measurements to the controlled input of the plant $u(t)$ given as:

$$K(s) = \frac{U(s)}{E(s)} \tag{8.1}$$

But we also know that each transfer function can also be represented as a state space formulation, and so $K(s)$ can also be represented as:

$$K(s) = \begin{array}{l} \dot{\vec{x}}_c(t) = A_c \cdot \vec{x}_c(t) + B_c \cdot \vec{e}\,(t) \\ \vec{u}\,(t) = C_c \cdot \vec{x}_c(t) + D_c \cdot \vec{e}\,(t) \end{array} \tag{8.2}$$

Here, subscript "c" denotes the controller or compensator in the state vector and state space matrices. If $d(t) = 0$ then measurements are directly fed into the control system to get the compensation commands i.e., $u(t)$. The simulation block diagram in Fig. 8.3 shows that a state space plant can be represented as an integrator with matrices as gains. The objectives are required outputs $y(t)$ with C as output gain matrix and D as input–output gain matrix or feedforward matrix. Measurements may be different than the actual required outputs; combinations of state vector and input vector to represent measurement are given by C_m and D_m matrices. A controller may be represented as a transfer function as well, such as a PID block in Fig. 8.3. PID is one of the most commonly used and technologically accepted controller design methods.

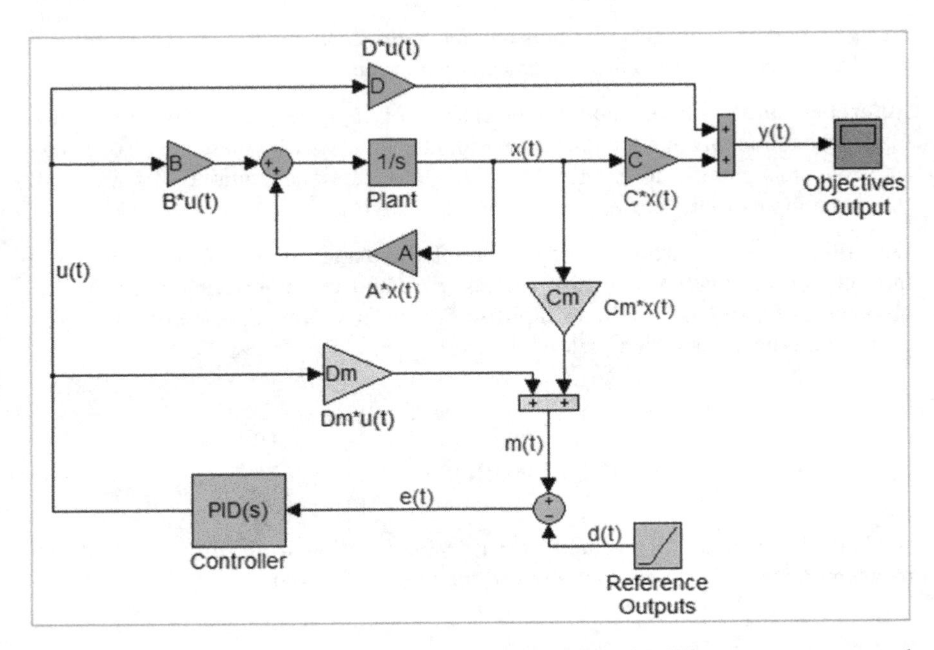

Fig. 8.3 Simulation diagram of control system as PID controller with different measurements and objectives. Actuators and sensor measurement transfer functions are taken as unity with no reference input

8.2 Error Model

In linear systems it is also very common to write model equations in terms of the error dynamics by defining state vector appropriately. A stable linear system will settle at zero steady-state if initial conditions are zero and there is no constant term being added. The error model shows that error goes to zero during the process. In this case zero is the desired reference value which can eliminate the signal $d(t)$ from the simulation. The error model of the plant dynamics are given with the same state space matrices while state variables are defined as the error in the states.

Example 8.1 Linear Model of an Inverted Pendulum

State space model of a simple pendulum (Example 2.1) is given as

$$\begin{aligned} \dot{\theta} &= \theta \\ \ddot{\theta} &= -\frac{g}{l}\sin\theta \end{aligned} \tag{8.3}$$

If we assume $\theta_{\text{ref}} = \frac{\pi}{2}$, write a state space formulation in the error dynamics

Solution Let

$$x_1 = \theta - \theta_{\text{ref}} \tag{8.4}$$

So

$$\dot{x}_1 = \dot{\theta} = x_2 \tag{8.5}$$

This assumes that the zero reference velocity is required at $\frac{\pi}{2}$.

So nonlinear state space equations will be given as

$$\begin{aligned} \dot{x}_1 &= x_2 \\ \dot{x}_2 &= \frac{g}{l} \cos x_1 \end{aligned} \tag{8.6}$$

Because $-\sin(\theta - \pi/2) = \cos(\theta)$.

Note that this is a nonlinear system, when we will linearize it then we get the same state space matrices because of its linearization points. In the case of nonlinear systems this problem does not arise and same state space matrices can be used with constant vector being added or subtracted. But for the error dynamics, equations with dynamical reference trajectories, this model needs to be addressed with some additional matrix algebra.

8.3 Estimator or Observer

A set of equations that estimates the state vector during process is called an estimator or observer. In digital control theory the word "estimator" is more popular or appropriate and in continuous time system "observer" is commonly used. In reality all states of the system are not explicitly available because taking measurements of each and every state through sensor is almost impossible. So in order to effectively take control of all states it is required to observe states through some mechanism. The first question arises that if we know model, initial conditions, and starting time as well as controlled input to plant we can simultaneously run this model to get the values of these states. True, these are called open loop observers as given in Fig. 8.4.

State space formulation for open loop observer is given as

$$\begin{aligned} \dot{\overrightarrow{x}}_e(t) &= A \cdot \overrightarrow{x_e}(t) + B \cdot u(t) \\ \overrightarrow{y_e}(t) &= C \cdot \overrightarrow{x_e}(t) + D \cdot u(t) \end{aligned} \tag{8.7}$$

Observer output $\overrightarrow{y_e}(t)$ should reach the actual output of plant $\vec{y}(t)$ simultaneously as both systems are starting at the same initial conditions as well as excited with the same input vector. In reality it is not possible due to the fact that actual plant, which also have uncertainties, facing environmental disturbances, approximation errors and transmission delays, etc. does not behave exactly as same as the analytical open

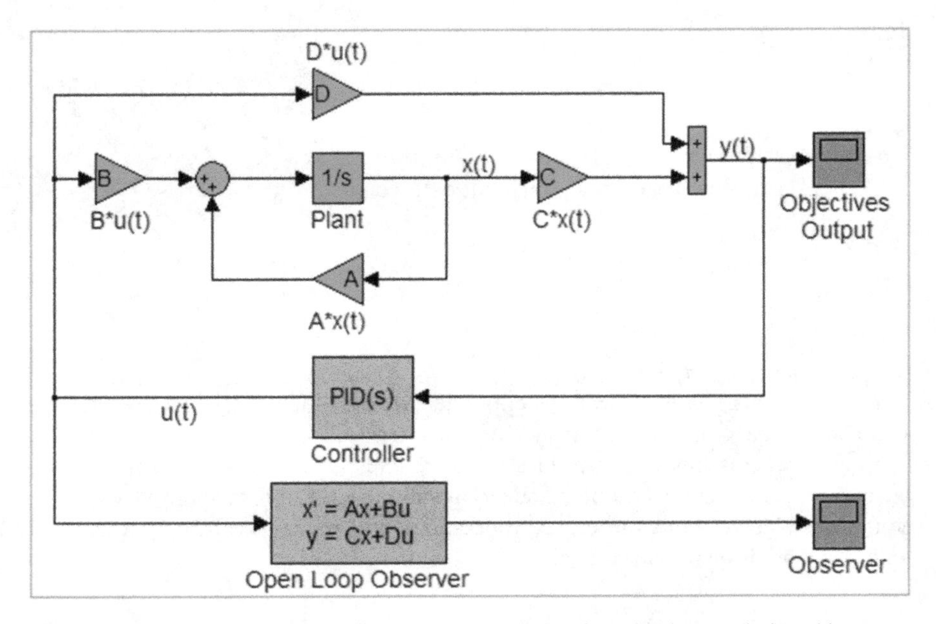

Fig. 8.4 Open loop observer to estimate the states of the plant simultaneously by taking same inputs and same state matrices

loop observer. A closed loop system is required to get the actual states of plant and passes to the observer, which estimates the correct values of states instantaneously for regulation or control commands. A closed loop observer needs measurements or error with respect to measurements and input calculated by controller to give the required estimates of the states.

A closed loop observer shown in Fig. 8.5 is called closed loop due to the feedback loop from the error between outputs of the plant and the observer. The task of the closed loop observer is to minimize the error between actual outputs of the plant and observed outputs and as $e(t) \rightarrow 0$ then $\vec{y}_o(t) \rightarrow \vec{y}(t)$, thus matching the observer with plant. The equation of the closed loop observer is given as

$$\begin{aligned} \vec{\dot{x}}_o(t) &= A \cdot \vec{x}_o(t) + B \cdot \vec{u}(t) + L\left\{\vec{y}(t) - \vec{y}_o(t)\right\} \\ \vec{y}_o(t) &= C \cdot \vec{x}_o(t) + D \cdot \vec{u}(t) \end{aligned} \tag{8.8}$$

An input vector $\vec{u}(t)$ may be composed of feedback controller output and reference input. The output of closed loop observer regulates the plant, which is actually a controller gain matrix multiplying either states or estimates of states. A control system is actually a combination of both closed loop estimator and regulator to provide controlled input. This is also referred to as compensator—a combination of regulator and estimator as given in Eq. (8.2) and below:

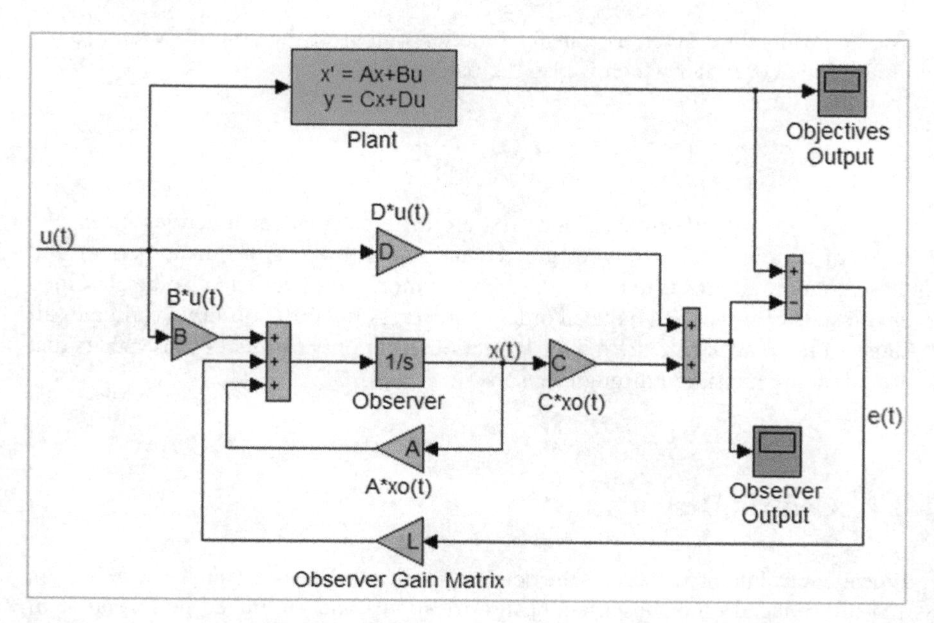

Fig. 8.5 A closed loop observer to estimate the real states by minimizing error between actual and estimated output with same input source *u(t)*

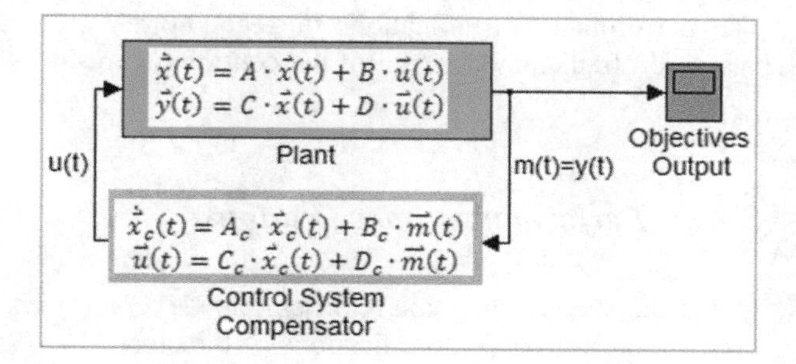

Fig. 8.6 Closed loop plant and control system compensator

$$\begin{aligned} \dot{\vec{x}}_c(t) &= A_c \cdot \vec{x}_c(t) + B_c \cdot \vec{m}(t) \\ \vec{u}(t) &= C_c \cdot \vec{x}_c(t) + D_c \cdot \vec{m}(t) \end{aligned} \tag{8.9}$$

The output of plant is measurements as an input to the control system (compensator) in case of no desired reference at output. Similarly, output of compensator is the input to the plant without any reference input to the plant. The control system or compensator matrices A_c, B_c, C_c and D_c are analytically computed using theories of linear algebra.

In this case it is clear that plant is represented as transfer function from output (measurements) to input given as:

$$G(s) = \frac{M(s)}{U(s)} \tag{8.10}$$

The controllers are given as transfer function output of the control system to the input of the controller which is also the output of plant

$$K(s) = \frac{U(s)}{M(s)} \tag{8.11}$$

It must be clear that both $K(s)$ and $G(s)$ are different transfer functions where the order of denominator is always greater and equal to the order of numerator and these are not reciprocal to each other. Furthermore, observer can also be classified as full state observer and reduced order observer. A full order observer estimates all states of the plant while the reduced order observer only estimates those states that are not being measured through sensors.

8.4 Control Design

Linear control theory describes the details of controller design, compensator design, and full/reduced order observers design for robust and optimized performance of the plant. A simple controller may be a PID controller which is designed to stabilize the plant with specific parameters. A PID controller is equivalent to lead-lag compensator design mathematically and both are basically second-order systems. The performance specifications of second-order systems are applicable to the designs in frequency domain for the control systems.

8.4.1 Performance Specifications

In this chapter, we will explore the simple PID control system design to understand the basic performance specifications of the closed loop system. Figure 8.7 shows the typical second-order transient response for a variable $y(t)$ to achieve the final desired value $y_f(t)$.

8.4.1.1 Steady-State Value

A value of response achieved as a final value after excitation is denoted by y_{ss}. A tolerance band is usually defined on final value and when a value is achieved within that band for significant time then it is called a steady-state value. In Fig. 8.7, a red line shows the final value y_f and as $t \to \infty$ then $y_{ss} \to y_f$. Two lines are shown as $\pm 10\,\%$ of y_f and response to $y(t)$ settles to y_{ss} even it still oscillates within $\pm 10\,\%$ tolerance limits.

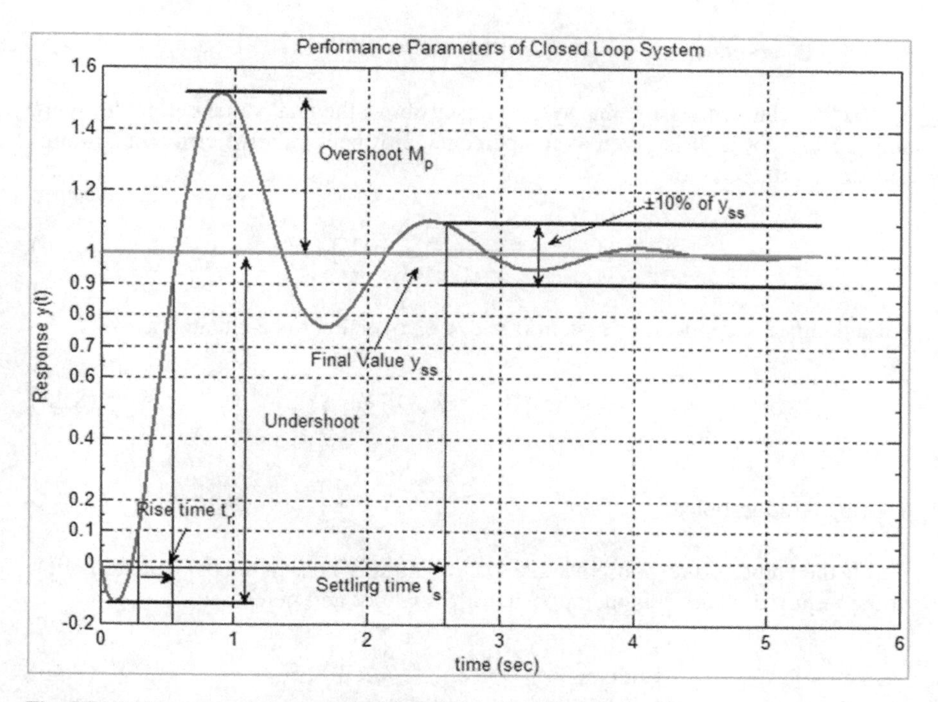

Fig. 8.7 Performance parameters of closed loop system in time domain

8.4.1.2 Settling Time

This is the time it takes for the response to decay and remain within the specified tolerance band of the steady-state value. It is denoted by t_s and the value is given when the response enters the tolerance band and remains in the band afterwards. An usual approximation of settling time for a second-order damped system is given in terms of damping frequency σ or damping ratio ζ and natural frequency ω_n.

$$t_s \approx \frac{4.6}{\sigma} = \frac{4.6}{\zeta\omega_n} \tag{8.12}$$

8.4.1.3 Rise Time

This is the time required to raise the response from 10 % of final value to 90 % of the final value. Some text give from 0 to 90 % of time but this can only be a true measure if there is no undershoot in the response. The rise time is approximated by using natural frequency of second-order system given as:

$$t_r \approx \frac{1.8}{\omega_n} \tag{8.13}$$

8.4.1.4 Overshoot

This is the relative measure that system raised above the final value before decaying and settling down. It is given as the percentage of peak relative error (maximum) and steady-state value

$$M_{\mathrm{p}} = \frac{(y_{\max} - y_{\mathrm{ss}})}{y_{\mathrm{ss}}} \times 100\% \tag{8.14}$$

For a damped second-order system $0 < \zeta < 1$, overshoot is estimated as

$$M_{\mathrm{p}} \approx \exp\left(-\pi\zeta/\sqrt{1-\zeta^2}\right) \tag{8.15}$$

8.4.1.5 Undershoot

This is the value of a response when a system initially goes in the opposite direction of the required value. It is the opposite of overshoot and given as:

$$\text{Undershoot} = \frac{(y_{\mathrm{ss}} - y_{\min})}{y_{\mathrm{ss}}} \times 100\ \% \tag{8.16}$$

In many cases undershoot does not occur, especially if all initial conditions are either positive or negative. The definition of responses can be a little different if a response starts from a higher value and settles at lower value. In MATLAB, a command **S=stepinfo(y,t,yfinal)** calculates the performance parameters of the required functions.

8.5 Controlling the Response

Designing a control system from a model is often a challenging task due to the variety of parameters to be satisfied. These parameters are related with physical abilities of the plant or the system under control. It may not be possible to achieve a very small rise time if the natural frequency of the system is high, and higher damping in the system increases settling time and overshoot, etc. It is difficult to fully control and change the natural characteristics of the plant through transfer functions, pole–zero cancellations, or lead-lag compensation. However, control objectives can be achieved significantly with appropriate control designs. There are different design methods to appropriately shape the dynamic response of the plant by feedback loop. A feedback function is designed to operate in a loop as given in different configurations in Figs. 8.1, 8.2, 8.3, 8.4, 8.5, and 8.6. Simplest of all approaches is a unity feedback with PID controller given in Fig. 8.3. The simplest and most commonly used controller design is known as the PID control design or the Proportional–Integral–Derivative control design. There can be other methodologies as well to design a controller in the feedback loop.

8.5.1 PID Control

In order to explain the Proportional–Integral and Derivative (PID) control methodology, we draw Fig. 8.3 with the transfer of a plant as well as a state space representation of a plant shown in Fig. 8.8.

PID controller itself is a single input to single output transfer function from a plant input to the error between plant output and desired reference given in three terms separately

$$u(t) = K_{\mathrm{p}} \cdot e(t) \tag{8.17}$$

$$u(t) = K_{\mathrm{I}} \cdot \int_0^t e(\tau)\mathrm{d}\tau \tag{8.18}$$

$$u(t) = K_{\mathrm{D}} \cdot \frac{\mathrm{d}e(t)}{\mathrm{dt}} \tag{8.19}$$

The parameters in the three terms are K_{p}, K_{I} and K_{D} as the design values in the PID controller. K_{p} is a simple proportionality between the output of PID and input to the PID $e(t)$ simply representing increasing or decreasing gain based upon magnitude of error. When error is integrated in order to observe settling of the response, then integral parameter K_{I} is tuned for appropriate gain. The sudden changes and rate of change of error is observed and tuned through derivative K_{D} control based upon error of the plant output. The PID controller gain is also given as:

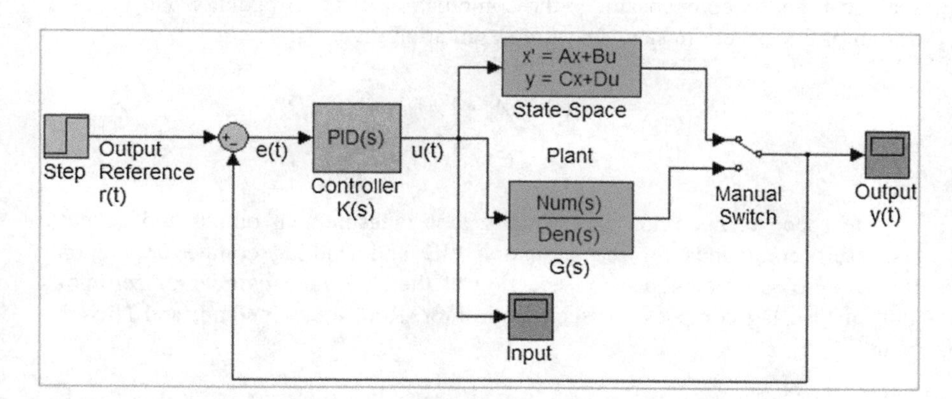

Fig. 8.8 Feedback control loop with unity feedback, output reference, and plant configuration in state space as well as equivalent transfer function

$$K(s) = \frac{U(s)}{E(s)} = K_{\mathrm{p}} + \frac{K_{\mathrm{I}}}{s} + K_{\mathrm{D}} \cdot s \tag{8.20}$$

This is the second-order transfer function required to solve for the numerator or denominator from this equation. This will be achieved by taking inverse Laplace transform term-by-term and then finding an actual transfer function. There is often a requirement to use the controller as either PI or PD due to simplification and lesser design specifications. The integral parameter K_{I} controls the steady-state error which may occur due to the proportional term K_{p}. Damping overcomes the overshoot, and the speed of the response regulates by the derivative parameter K_{D}.

8.5.2 Compensation

Compensator is a simple transfer function from an input of the plant to an output of the plant error given in the pole zero gain representation

$$K(s) = K_{\mathrm{c}} \frac{s + z_{\mathrm{c}}}{s + p_{\mathrm{c}}} \tag{8.21}$$

The subscript "c" is for compensator and if $z_{\mathrm{c}} < p_{\mathrm{c}}$ then this is known as lead compensator. Lead compensator is equivalent to the derivative controller term K_{D}, which is used to decrease the overshoot of the response and increase bandwidth or damping in the system. If the pole of the compensator is lagging the zero of compensator $z_{\mathrm{c}} > p_{\mathrm{c}}$ then the system is known as lag compensator, which is equivalent to the integral control term K_{I} whose primary objective is to decrease the steady-state error. A second-order compensator is actually equivalent to the PID control design, which may have both lead and lag properties for full design specifications. A compensator is the combination of both regulator and observer design of the system in state space representation given as

$$K(s) = \begin{array}{l} \dot{\vec{x}}_{\mathrm{c}}(t) = A_{\mathrm{c}} \cdot \vec{x}_{\mathrm{c}}(t) + B_{\mathrm{c}} \cdot \vec{m}(t) \\ \vec{u}(t) = C_{\mathrm{c}} \cdot \vec{x}_{\mathrm{c}}(t) + D_{\mathrm{c}} \cdot \vec{m}(t) \end{array} \tag{8.22}$$

A state space compensator can directly take measurement output and generate regulating commands for the actuators. PID and lead-lag compensator adopts frequency-based techniques for selection of the PID gain matrices or zero/pole/ gain of lead-lag compensator such Bode Plot, Root locus method, and Nicholas chart.

8.5.3 Pole Placement Control Design

Regulator and estimator are designed in a state space formulation in order to change the closed loop poles of the system. A regulator is achieved by designing a matrix K taking state input to regulate the feedback command

$$\vec{u}(t) = -K \cdot \vec{x}(t) \tag{8.23}$$

The matrix K is of order $p \times n$ for p inputs and n states and it is selected in such a way that closed loop poles of the system are placed at the desired locations. The closed loop poles are active when the feedback command is regulating the system and the dynamic profile of the system is determined by the location of the poles. In Eq. (8.22), $C_c = -K$ and $D_c = 0$ for simple regulator design of the closed loop system, where $B_c = L$ is an observer gain matrix as shown in Fig. 8.5. The observer gain matrix is chosen such that closed loop poles for observer are stable and faster than regulator poles. The matrix A_c is a feedback matrix from both regulator and observer poles.

8.6 Shaping the Dynamic response

The dynamics response can be shaped either for selection of PID gains, or lead-lag compensator using MATLAB optimization tools. The performance specification met using different methodologies of designs. Problem 8.6 describe the simulation technique in detail for pole placement and shaping the dynamic response accordingly.

Example 8.2 PID Tuning
Simulate the open loop system and tune the PID gains in order to obtain the desired characteristics as given in Fig. 8.8 using step input.

$$\begin{gathered}\dot{\vec{x}}(t) = \begin{bmatrix} -1 & 4 \\ -4 & -1 \end{bmatrix} \cdot \vec{x}(t) + \begin{bmatrix} 1 \\ 3 \end{bmatrix} \cdot \vec{u}(t) \\ y(t) = \begin{bmatrix} 2 & -2 \end{bmatrix} \cdot \vec{x}(t) \\ t_r < 800\,\text{ms} \quad t_s < 4\,\text{s for} \pm 10\,\% \text{ of } y_{ss} \quad \text{Over Shoot } < 9\,\% \end{gathered} \tag{8.24}$$

Get the information by using stepinfo command and plot the open loop and closed loop response. Is it first peak in the profile which is used to measure overshoot?

Solution The open loop response of this system given in Fig. 8.7 with following parameters

```
S=stepinfo(y.signals.values,y.
time,1.65,'SettlingTimeThreshold',0.1)
RiseTime: 0.2535
```

```
SettlingTime: 2.9753
SettlingMin: 1.2576
SettlingMax: 2.5016
Overshoot: 51.611
Undershoot: 13.766
Peak: 2.5016
PeakTime: 1.9015
```

PID block tunes itself for the different values and options and after tuning to step input response in order to meet the desired objective, the following parameters are achieved for $y_{ss} = 1$ as shown in Fig. 8.9.

```
RiseTime: 0.71328
SettlingTime: 3.5662
SettlingMin: 0.7687
SettlingMax: 1.063
Overshoot: 6.3006
Undershoot: 1.8029
Peak: 1.063
PeakTime: 4.0859
```

The tuned response for PID parameters is $K_p = 0 \quad K_I = 0.5618 \quad K_D = 0$. It is obvious that overshoot calculates from second peak, which gives the maximum value of $y(t)$. It is to be noted that the open loop system is internally stable, and it is more difficult to achieve control parameters for internally stable systems as compared to internally unstable systems in general.

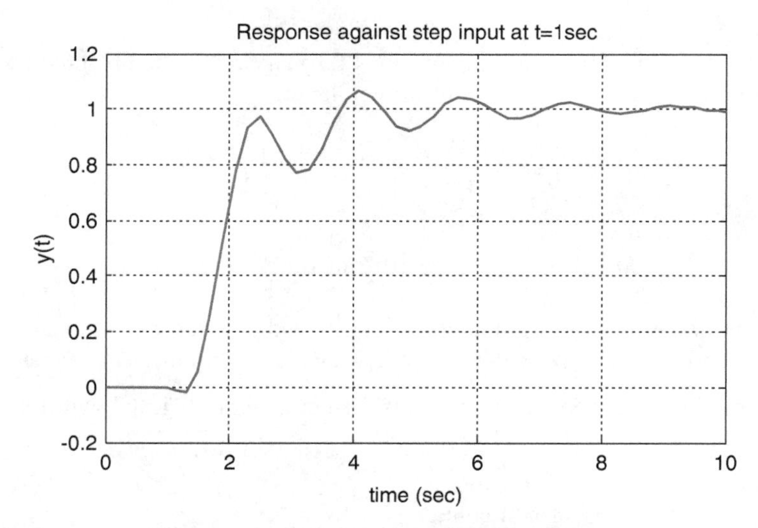

Fig. 8.9 Step response of a system with tuned PID controller starting at t=1 s

8.7 Gain Scheduling in Multivariable Control

Controlling more than one output using one or more inputs makes a multivariable control problem. Each output–input loop is treated differently and some loops may be ignored due to insignificant contributions in the loop. The input to the plant is dependent upon the combination of different loops. The gains in each loop need to be adjusted so that the overall system remains bounded and stable.

Example 8.3 Tune a PID controller using signal constraints block of the following system:

$$\vec{\dot{x}}(t) = \begin{bmatrix} 2 & 3 \\ -1 & 3 \end{bmatrix} \cdot \vec{x}(t) + \begin{bmatrix} 0 \\ 2 \end{bmatrix} \cdot \vec{u}(t)$$

$$\vec{y}(t) = \begin{bmatrix} 1 & 0 \\ 1 & 2 \end{bmatrix} \cdot \vec{x}(t)$$

Solution This is a multivariable system due to two outputs, and PID is a SISO system so both outputs tune separately. The gain multiplies in both loops in order to effectively control the plant. The signal constraint can also optimize the both gains g_1 & g_2 for PID controllers $K_1(s)$ & $K_2(s)$ respectively for components of same input. It is important to note that PID blocks $K_1(s)$ & $K_2(s)$ are tuned independently to meet performance specifications before tuning of gains g_1 & g_2 for input. The inputs $u(t)$ are given as:

$$u(t) = g_1 \cdot u_1(t) + g_2 \cdot u_2(t)$$

The signal constraint block optimizes for performance specification of net input to the plant. This signal constraint block optimizes outputs as well, and in many cases it can also optimize more than one variable at a time (Fig. 8.10).

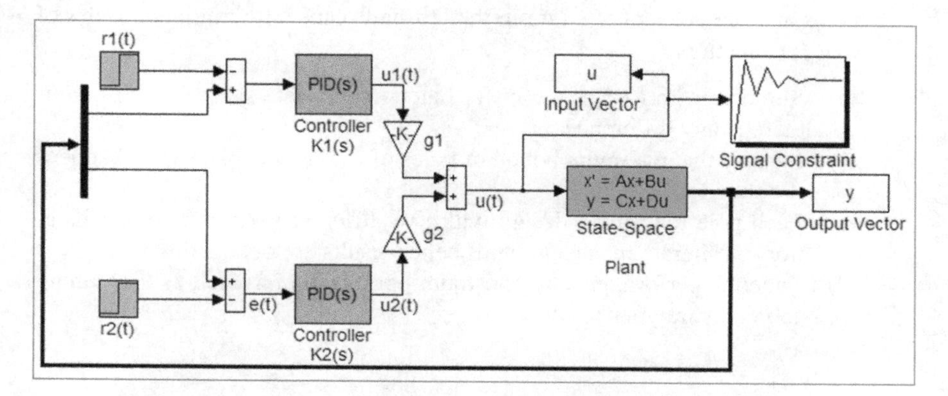

Fig. 8.10 Simulation block with two PID loops and gain scheduling for input

Problems

P8.1 Draw the block diagram of the systems given in Examples 8.1 and 8.2 by treating each element differently. There will be n-integrators for n states and each element of matrices will be a gain in the loop.

P8.2 In Fig. 8.2, if we augment the actuators and measurement into the plant state space, what will happen to the order of plant, which matrices will be changed and how this will affect control loop? Take the following transfer functions and augment with the plant of Example 8.1.

$$\mathrm{Act}(s) = \frac{1}{s + 1000} \quad \mathrm{Sensor} = \frac{2}{s + 2000}$$

P8.3 Find a PID gains for the system in Problem 8.2 with the same performance specifications given in Example 8.2.

P8.4 Simulate the open loop observer of the problem in Example 8.1 and note the error. Introduce the small random noise in feedback path with noise power 0.0001 W and simulate again to observe changes in error.

P8.5 Tune the block of Problem 8.4 for different rise time and overshoot of the input. What difficulties are encountered in order to tune PID blocks and input optimization? What are the minimum rise time, peak amplitude, and steady-state error achieved in your solution? Is the solution unique?

P8.6 **Pole Placement using Transfer function**: Simulate Fig. 8.6 for the following systems:

1. Example 5.5
2. System in Fig. 5.28
3. Problem 7.6
4. Example 8.1
5. Example 8.2

Choose $K(s)$ appropriately such that the zeros of $K(s)$ cancel all the unstable poles and the poles of $K(s)$ are in the left half plane with minimum order of transfer function.

(a) Simulate with $K(s)$ as a transfer function and state space and observe if there is any difference.

(b) Identify the maximum bound of poles of $K(s)$ and explain what happens if bound is violated.

(c) Each pole placement design will give different performance specifications; re-iterate the design until better results are obtained.

(d) Tune the performance by constraint optimization as well as PID tuning and compare your results.

References

1. Friedland, Bernard. 2005. *Control System Design—An Introduction to State Space Methods*. Mineola, NY: Dover Publications.
2. Chen, Chi-Tsong. 1999. *Linear System Theory and Design*, 3rd ed. New York: Oxford University Press.
3. Heij, Christiaan, André Ran, and Freek van Schagen. 2007. *Introduction to Mathematical Systems Theory—Linear Systems, Identification and Control*. Basel: Birkhäuser Verlag.
4. Hespanha, Joao. 2009. *Linear Systems Theory*. Princeton, NJ: Princeton University Press.

Chapter 9
Recent Applications of Bond Graph Modeling

Bond graph modeling is used in diverse engineering applications due to its universality between energy systems. Every year the number of research papers and publications in the domain of modeling and engineering using this technique is on rise. Lagrangian modeling is a popular method, especially in robotic manipulators, but with the help of bond graphs these models can be extended with electromechanical systems of actuators and sensors. In the last decade, numerous papers in conferences and journals have appeared for bond graph modeling and simulation. It is being taught at undergraduate and graduate level courses in different universities showing the developing interest in this technique. Bioengineering or biomedical engineering is a key area of interest for research and development for its applications. Bond graph modeling made its way to this area by providing an alternative paradigm for modeling and simulation. In this chapter, we will focus on the recent applications of bond graph in biomedical engineering followed by some applications in a few other fields.

9.1 Bond Graph Models of the Physiological Elements

Physiological systems have been the subject of interest to a variety of individuals including artists, physicians, athletes, scientists and researchers. A primary object of interest in the physiological system is the muscle structure; researchers have studied muscles from varying perspectives and developed multidimensional understandings of its structure. Early biomechanics researchers understood the muscle as a spring-like element and modeled it as stiff object; however, it is now understood as a combination of a contractile and a pair of stiff elements in series or parallel arrangement. This model has been studied extensively in the past 50 years and still solicits interest of researchers and physiologists worldwide, who either use a reductionist approach or synergist methods. Complexities have been added and studied for this model, and various other structures have also been proposed.

A.M. Mughal, *Real Time Modeling, Simulation and Control of Dynamical Systems*,
DOI 10.1007/978-3-319-33906-1_9

Muscular analysis involves mechanical, chemical, biological, and electrical properties of the muscle structures. Bond graph modeling provides suitable tools when dealing with the systems involving different kinds of energy flows. Bond graph model of a system deals with the flow of power from one node to another; it uses generalized power and energy variables to track the flow of power. The use of bond graphs provides an alternative approach in physiological modeling. This approach is useful when determining efforts (Forces) or flows (velocities) at different nodes (joints, sections) for musculoskeletal systems. This technique involves simple additions of either efforts (forces) or flow (velocities) at the junctions and can also be used by physiologists not familiar with mechanical or electrical modeling techniques.

9.1.1 Muscular Structures

Hill-type muscle models are used by physiologists and researchers for the biomechanical modeling and development of the musculoskeletal control strategies. Muscle structures exhibit mechanical, electrical, and chemical properties simultaneously. Figure 9.1 represents the block diagram of two Hill-type muscle models; their respective bond graphs are given in Figs. 9.2 and 9.3, generated with 20-Sim software.

In these models, we represent general compliance with C, which is reciprocal of stiffness k, in series and parallel elements component C_{s} and C_{p} respectively. The mass of muscle m is represented by I (inertial component), and viscous damping of contractile element is represented by B. The source of effort of force acting on the muscle is S_{e}, which is externally applied to the muscle. The state variables in Hill's 1st muscle models are generalized displacement (q) at port 3 and port 5, and momentum (p) at port 2. Bond graph can be used to write linear as well as nonlinear state equations. Equation (9.1) represents the state equation of the 1st Hill model.

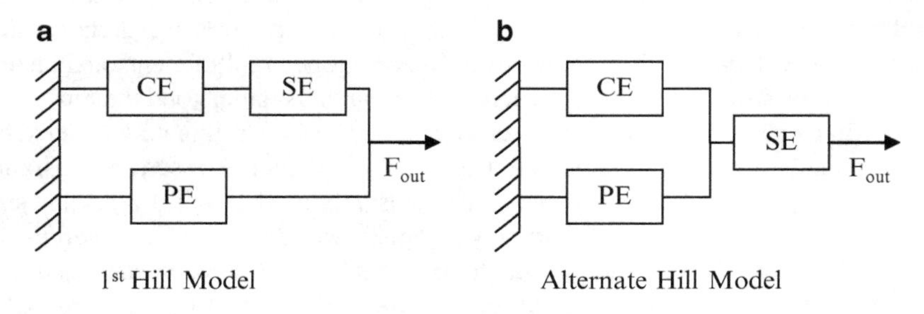

Fig. 9.1 An example of two Hill-type muscle models

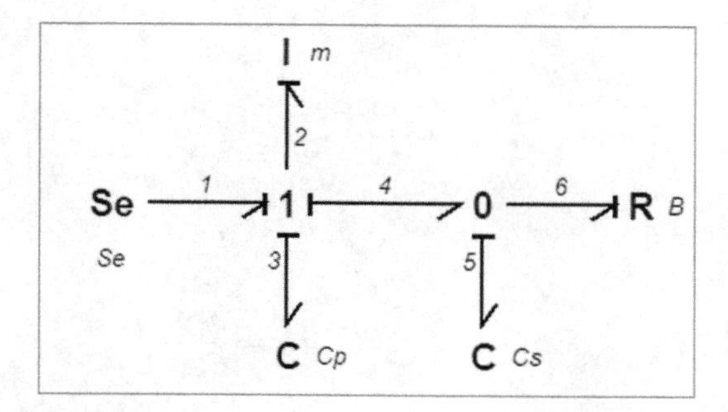

Fig. 9.2 Bond graph of 1st Hill muscle model

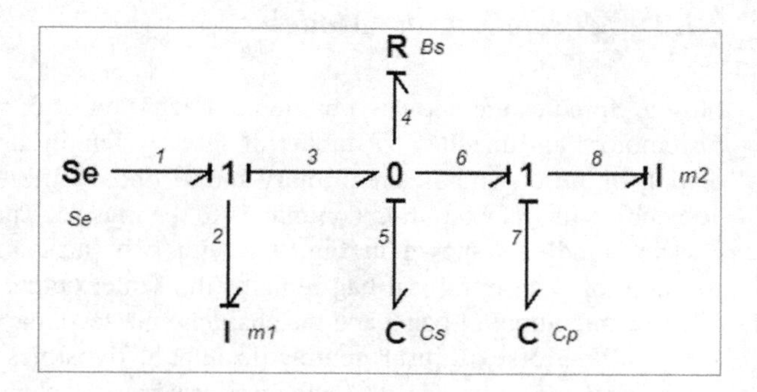

Fig. 9.3 Bond graph of an alternate Hill muscle model

$$\begin{aligned}\dot{p}_2 &= S_E - \frac{q_3}{C_p} - \frac{q_5}{C_s} \\ \dot{q}_3 &= \frac{p_2}{m} \\ \dot{q}_5 &= \frac{p_2}{m} - \frac{1}{B}\left(\frac{q_5}{C_s}\right)\end{aligned} \tag{9.1}$$

The output force of the 1st model is given by Eq. (9.2)

$$F_{out} = \frac{q_5}{C_s} \tag{9.2}$$

Similarly, we can write state equations for alternate Hill muscle model. In this model, there are two inertial components m_1 and m_2 at both ends of muscles with a total of four state variables. These variables are generalized momentums at port 2 and 8 and generalized displacement at port 5 and 7; output is the force at port 8, which is the rate of change of momentum at port 8.

$$
\begin{aligned}
\dot{p}_2 &= S_E - \frac{q_5}{C_s} \\
\dot{q}_5 &= \frac{P_2}{m_1} - \frac{P_2}{m_1} - \frac{1}{B_s}\left(\frac{q_5}{C_s}\right) \\
\dot{q}_7 &= \frac{P_8}{m_2} \\
\dot{p}_8 &= \frac{q_5}{C_s} - \frac{q_7}{C_p}
\end{aligned}
\tag{9.3}
$$

9.1.2 Muscle Spindle Model

Muscle Spindles are activated by neural excitation of dynamic (γ_d) and static (γ_s) fusiomotors and mechanical inputs of fascicle length and fascicle velocity. The outputs of muscle spindle are primary and secondary afferents, which modulate the descending neural excitation commands to the muscle. The model structure for the muscle spindle is shown in Fig. 9.4 with two intrafusal bags and chain. The structure of each intrafusal bag remains the same; output of bag 1 is the primary afferent and output of bag 2 and the chain are the secondary afferent. These outputs are functions of extrafusal contractile length, dynamics of intrafusal fiber, and mechanical and neural inputs. Each intrafusal bag and chain consists of a contractile element and series and parallel stiffness (spring) elements. Mechanical inputs are the fascicle length and velocity from musculoskeletal structure. The neural outputs, primary (*Ia*) and secondary afferent (*II*) from the muscle spindle add to neural excitation of the muscle. The general structure of the intrafusal bags and a chain are given in Fig. 9.4. The components of muscle spindle and sensory region (SR) are shown in Fig. 9.5. It consists of a parallel viscous element and a stiffness (spring) element, while the neural input is represented by Γ. These elements may be treated as linear as well as nonlinear elements, and a combination of several terms. Contractile elements consist of fibers, which can be modeled separately or combined into an equivalent effect.

The components shown in this model in Fig. 9.5 are a mechanical source of effort (S_e), and fusiomotor input Γ as modulated source of effort, series and parallel compliances (stiffness), viscous damping component and mass. State equations of

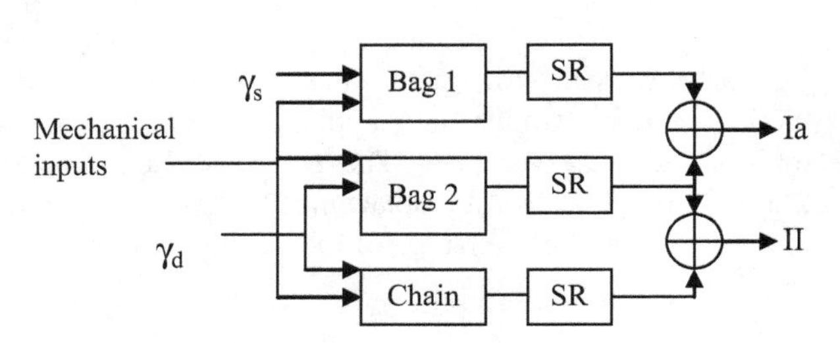

Fig. 9.4 General structure of the muscle spindle receptor

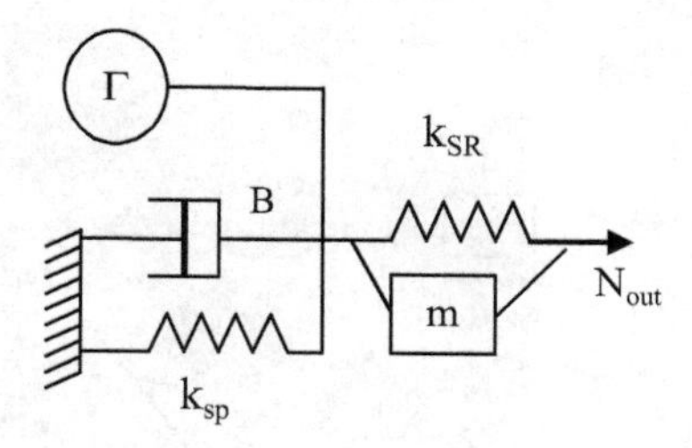

Fig. 9.5 Structure of intrafusal bag and chains with sensory region (SR)

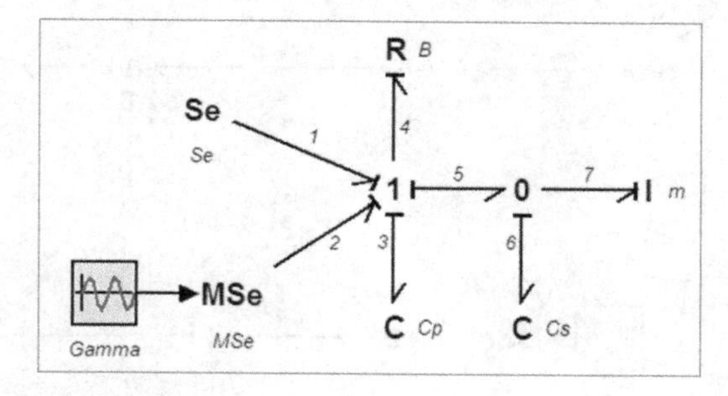

Fig. 9.6 Bond graph model of muscle spindle

an intrafusal bag model with sensory region bag are given in Eq. (9.4) with three state variables, generalized displacement at port 3 and 6, and generalized momentum at port 7 (Fig. 9.6).

$$\begin{aligned}
\dot{q}_3 &= \frac{1}{B}\left(S_E + \Gamma - \frac{q_6}{C_s} - \frac{q_3}{C_p}\right)\\
\dot{q}_6 &= \frac{1}{B}\left(S_E + \Gamma - \frac{q_6}{C_s} - \frac{q_3}{C_p}\right) - \frac{p_7}{m}\\
\dot{p}_7 &= \frac{q_6}{C_s}
\end{aligned} \tag{9.4}$$

The combined effect of the bags and chain are shown in Fig. 9.7 for the model represented in Fig. 9.4. In this model, mechanical input S_e is from muscle force for fascicle length and velocity; the neural or excitation inputs, i.e., dynamic (γ_d) and static (γ_s) fusiomotor inputs are represented by the active bonds in this model via modulated source of efforts.

$$\dot{q}_{11} = \frac{1}{B_1}\left(S_e + \gamma_s - \frac{q_{11}}{C_p} - \frac{q_{17}}{C_s}\right) \tag{9.5}$$

$$\dot{q}_{12} = \frac{1}{B_2}\left(S_e + \gamma_d - \frac{q_{12}}{C_p} - \frac{q_{18}}{C_s}\right) \tag{9.6}$$

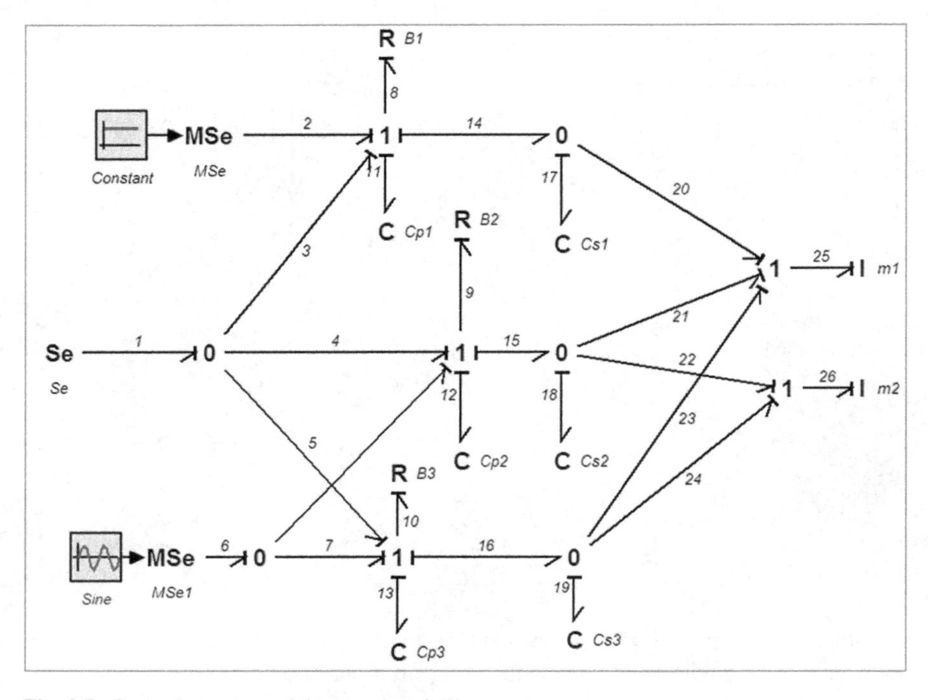

Fig. 9.7 General structure of the muscle spindle receptor

$$\dot{q}_{13} = \frac{1}{B_3}\left(S_e + \gamma_d - \frac{q_{13}}{C_p} - \frac{q_{19}}{C_s}\right) \tag{9.7}$$

$$\dot{q}_{17} = \frac{1}{B_1}\left(S_e + \gamma_s - \frac{q_{11}}{C_p} - \frac{q_{17}}{C_s}\right) + \frac{p_{25}}{m_1} \tag{9.8}$$

$$\dot{q}_{18} = \frac{1}{B_2}\left(S_e + \gamma_d - \frac{q_{12}}{C_p} - \frac{q_{18}}{C_s}\right) - \frac{p_{25}}{m_1} - \frac{p_{26}}{m_2} \tag{9.9}$$

$$\dot{q}_{19} = \frac{1}{B_3}\left(S_e + \gamma_d - \frac{q_{13}}{C_p} - \frac{q_{19}}{C_s}\right) - \frac{p_{25}}{m_1} - \frac{p_{26}}{m_2} \tag{9.10}$$

$$\dot{p}_{25} = \frac{q_{17}}{C_s} + \frac{q_{18}}{C_s} + \frac{q_{19}}{C_s} \tag{9.11}$$

$$\dot{p}_{26} = \frac{q_{18}}{C_s} + \frac{q_{19}}{C_s} \tag{9.12}$$

Two neural outputs, i.e., primary and secondary afferent, are given as:

$$F_{\text{out}\,1}(Ia) = \frac{1}{C_s}(q_{17} + q_{18} + q_{19}) \tag{9.13}$$

$$F_{\text{out}\,2}(II) = \frac{1}{C_s}(q_{18} + q_{19}) \tag{9.14}$$

9.1.3 Golgi Tendon Organ Model

Golgi Tendon Organ (GTO) responds to tensions in the muscle fibers and sends proprioceptive feedback to the muscle for activation. GTO is usually represented by static gains, but in bond graph, detailed models of GTO can be implemented. In a simple approach, GTO can be modeled as only an elastic spring element or combination of elastic element with viscous damping of negligible mass. In Fig. 9.8 S_e is the source of force from muscle with viscous and damping elements.

This is the simplest one-degree model of GTO. The output equation for this model is given as

$$\dot{x}_{\text{gto}} = \frac{1}{B_{\text{gto}}}\left(S_e - k_{\text{gto}} \cdot x_{\text{gto}}\right) \tag{9.15}$$

9.1.4 The Inertial Subsystem

A single-link inertial inverted-pendulum rigid-body-type biomechanical model for postural stability analysis is shown in Fig. 9.9. A single link with a triangular base

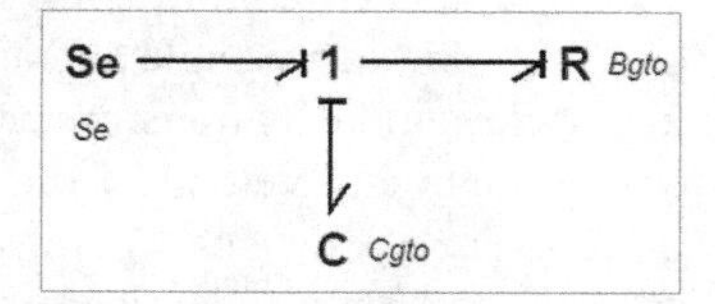

Fig. 9.8 Bond graph model of GTO

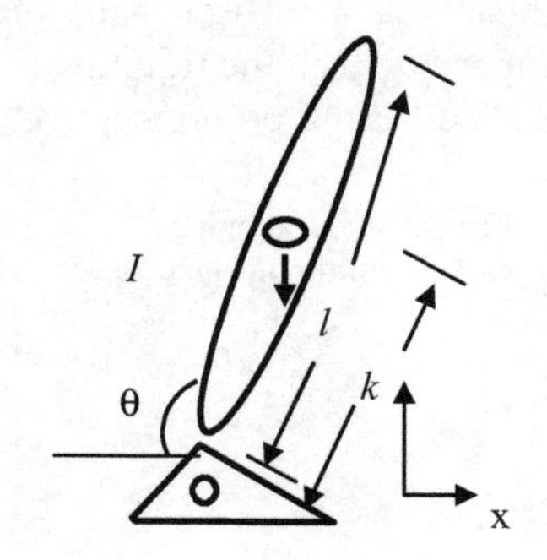

Fig. 9.9 Single-link inertial model for postural stability

of support, with mass m, inertia I, and a distance k from the center of mass to a joint was discussed in [1]. The inertial model has joint angle θ and angular velocity $\dot{\theta}$ as state variables for a 2nd order nonlinear system which was linearized at stable equilibrium (upright) position, and the dynamic state space representation of the inertial subsystem is given as

$$\ddot{\theta} = \frac{(\tau + mgk \cdot \sin\theta)}{(I + mk^2)} \tag{9.16}$$

The input torque to the inertial model comprises muscle force transformed into joint torque. The outputs of this model are angle and velocity that are transformed into fascicle length and fascicle velocity as inputs to the muscle spindle structure. The bond graph of an inverted-pendulum model has been discussed in detail in several textbooks.

9.2 The Complete Musculoskeletal System Model

The complete musculoskeletal systems with inertia, muscle, muscle spindle and GTO can be modeled using bond graphs. In this model we also need two-port elements called transformers for conversion from translational to rotational motion or vice versa. Inertial equations of joint mechanics are usually represented in angles, i.e., flexion of limbs (segments) at joints. Muscle equations are mechanical translational equations, so a transformer is needed at every joint to convert muscle forces into joints torques. Figure 9.10 shows the block diagram representation of a system with comprising components.

In Fig. 9.10, muscle output force is transformed into torque for inertial block and added up to external torque for limb movement. Transformer ratio (r) is the length of a limb. Similarly angle and angular velocity outputs from inertial block are sent to muscle spindle by transforming these to fascicle length and fascicle velocity. The output angle and velocity from inertial block are represented by the vector bond graphs. The vector bond graphs are used when multiple variables are connecting one port to another identically. These are also used for representing assembly of subsystems. The negative ratio (r) is used here because if force to moment is positive then the resulting angular motion results in muscle shortening. The muscle block takes proprioceptive feedback and produces appropriate forces for stable

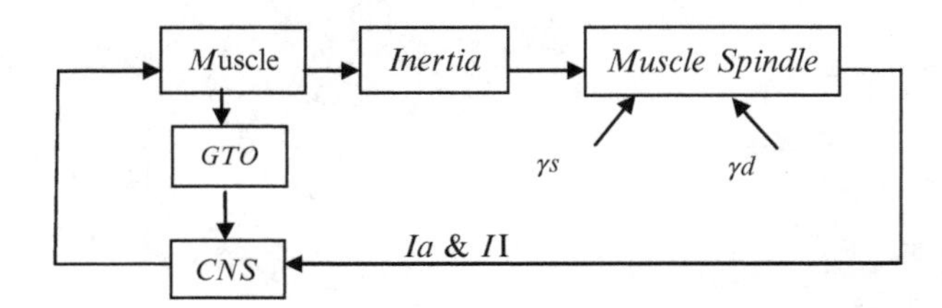

Fig. 9.10 Bond graph model of musculoskeletal system

movement. The muscle block has two components, one connected to central nervous systems (CNS) and the other producing controlled forces to inertial block. GTO also responds to the muscle forces and their feedback changes the neural excitation. The muscle output forces are transformed into joint torques adding up with external or disturbance torques and then cause the movement in inertial block. The musculoskeletal feedback system can be combined into a single state space representation for analysis and robust design of a controller structure. This dynamic system takes effect of all physiological variables interacting with each other in the closed loop response. The state variables of closed loop system are x_{M}, x_{S}, x_{G}, and x_{I} which represent the dynamics of muscle, muscle spindle, GTO and inertia respectively. The combined musculoskeletal model for the closed loop feedback system is given in Eq. (9.17).

$$\begin{bmatrix} \dot{x}_{\mathrm{M}} \\ \dot{x}_{\mathrm{S}} \\ \dot{x}_{\mathrm{G}} \\ \dot{x}_{\mathrm{I}} \end{bmatrix} = \begin{bmatrix} A_{\mathrm{M}} & A_{\mathrm{MS}} & A_{\mathrm{MG}} & 0 \\ 0 & A_{\mathrm{S}} & 0 & A_{\mathrm{SI}} \\ A_{\mathrm{GM}} & 0 & A_{\mathrm{G}} & A_{\mathrm{GI}} \\ A_{\mathrm{IM}} & 0 & 0 & A_{\mathrm{I}} \end{bmatrix} \cdot \begin{bmatrix} x_{\mathrm{M}} \\ x_{\mathrm{S}} \\ x_{\mathrm{G}} \\ x_{\mathrm{I}} \end{bmatrix} + \begin{bmatrix} B_{\mathrm{M}} \\ 0 \\ 0 \\ 0 \end{bmatrix} \cdot U \tag{9.17}$$

The order of the system is dependent upon the order of each subsystem. The muscle structure is 4th order, inertia is 2nd order systems, GTO is 1st order system, and muscle spindle is 6th order when using two intrafusal bags and a chain system represented in Fig. 9.4. The musculoskeletal system shown in Fig. 9.10 is 13th order system. However, if a simplified 3rd order muscle spindle structure is used then the overall system is reduced to 10th order. Only the muscle structure is directly affected by the input through CNS, and so in B matrix, there are zeros for other subsystems. The interaction matrices define the relationship between subsystems: A_{MS}, A_{MG}, represent interaction between muscle structure with muscle spindle and GTO respectively. Similarly A_{IM}, A_{SI}, A_{GI} represent the interaction of inertial subsystem with muscle, muscle spindle, and GTO. Transformer actions from muscle to inertial block and from inertial block to muscle spindle block are represented by T. There are three exogenous or disturbance inputs in the system, external (disturbances) torque added to the control torque, neural static (γ_{s}), and dynamic (γ_{d}) fusiomotor inputs to the muscle spindle. Figure 9.11 presents the complete musculoskeletal system with Hill's 1st and alternate muscle models respectively. In this case there is a differential causality at series compliance C_{s} at a zero junction, which adds one redundancy in the system, i.e., one dependent equation.

Figure 9.12 shows the complete musculoskeletal bond graph with Hill's alternate muscle model and this doesn't have any differential causality. The nonlinear model is given in Eqs. (9.18), (9.19), (9.20), (9.21), (9.22), (9.23), (9.24), (9.25), (9.26), and (9.27) for the 10th order system. In this system there is only one fusiomotor input with simple 3rd order muscle spindle model of, 2nd order nonlinear inertial model, 1st order GTO model and 4th order muscle model. This is an open loop model, which requires a feedback controller to regulate the sensory commands.

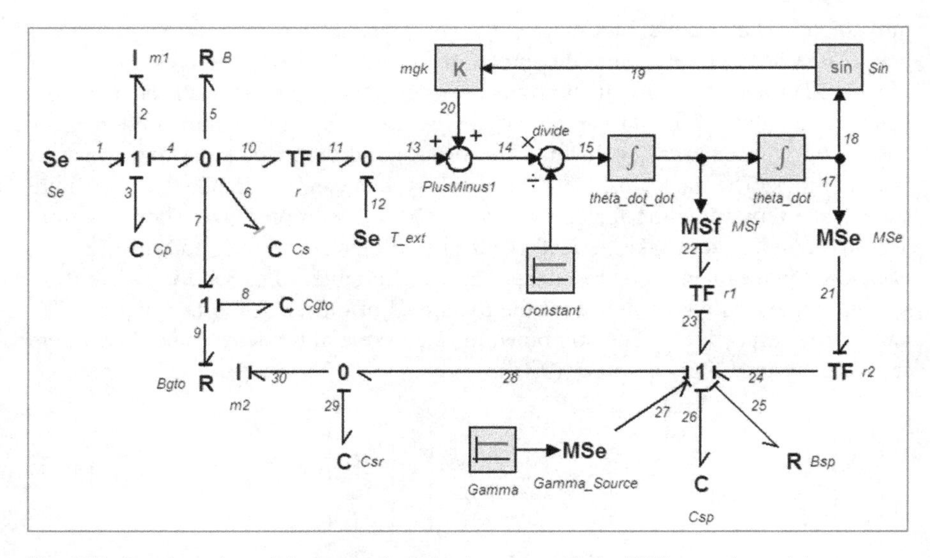

Fig. 9.11 Bond graph model of musculoskeletal system with 1st Hill's muscle model

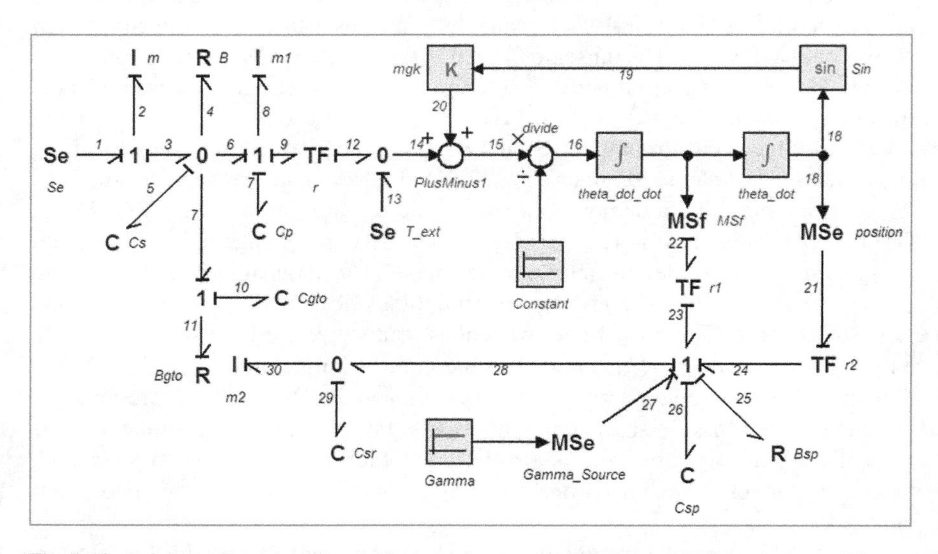

Fig. 9.12 Bond graph model of musculoskeletal system with alternate Hill's muscle model

$$\dot{p}_2 = S_\mathrm{e} - \frac{q_5}{C_\mathrm{s}} \tag{9.18}$$

$$\dot{q}_5 = \frac{p_2}{m} - \left(\frac{1}{B} - \frac{1}{B_\mathrm{gto}}\right) \cdot \frac{q_5}{C_\mathrm{s}} - \frac{1}{B_\mathrm{gto}} \cdot \frac{q_{10}}{C_\mathrm{gto}} \tag{9.19}$$

$$\dot{q}_7 = \frac{p_8}{m_1} \tag{9.20}$$

$$\dot{p}_8 = \frac{q_5}{C_s} - \frac{q_7}{C_p} + \frac{1}{l} \cdot \tau_{ext} \tag{9.21}$$

$$\dot{q}_{10} = \frac{1}{B_{gto}} \cdot \left(\frac{q_5}{C_s} - \frac{q_{10}}{C_{gto}} \right) \tag{9.22}$$

$$\dot{x}_{17} = \frac{1}{(I + m \cdot k^2)} \cdot \left(m \cdot g \cdot k \cdot \sin(x_{18}) + l \frac{p_8}{m_1} \right) \tag{9.23}$$

$$\dot{x}_{18} = x_{17} \tag{9.24}$$

$$\dot{q}_{26} = \frac{1}{B_{sp}} \cdot \left(\Gamma + \frac{1}{r_1} \cdot x_{17} + \frac{1}{r_2} \cdot x_{18} - \frac{q_{25}}{C_{sp}} - \frac{q_{29}}{C_{sr}} \right) \tag{9.25}$$

$$\dot{q}_{29} = \frac{1}{B_{sp}} \cdot \left(\Gamma + \frac{1}{r_1} \cdot x_{17} + \frac{1}{r_2} \cdot x_{18} - \frac{q_{25}}{C_{sp}} - \frac{q_{29}}{C_{sr}} \right) - \frac{p_{30}}{m_2} \tag{9.26}$$

$$\dot{p}_{30} = \frac{q_{29}}{C_{sr}} \tag{9.27}$$

9.2.1 The Extended Musculoskeletal Structure

The musculoskeletal structure augmented with a simple muscle spindle model is given in Figs. 9.11 and 9.12, with Hill 1st and alternate muscle models respectively. This model generates 10th order state space system given in Eqs. (9.18), (9.19), (9.20), (9.21), (9.22), (9.23), (9.24), (9.25), (9.26), and (9.27) which is an open loop system for sensory feedback to the CNS. This system can be augmented with the detailed model of muscle spindle receptor given in Fig. 9.7 with the structure given in Fig. 9.10. This system receives two inputs for fusiomotors instead of one static input. The overall combination is given in Fig. 9.13 with augmentation of control loop for the system. A simple PID tuner can tune the required loop with small noises. The detailed control analysis of the structures with static and dynamic inputs and disturbances were presented in research papers [2–5].

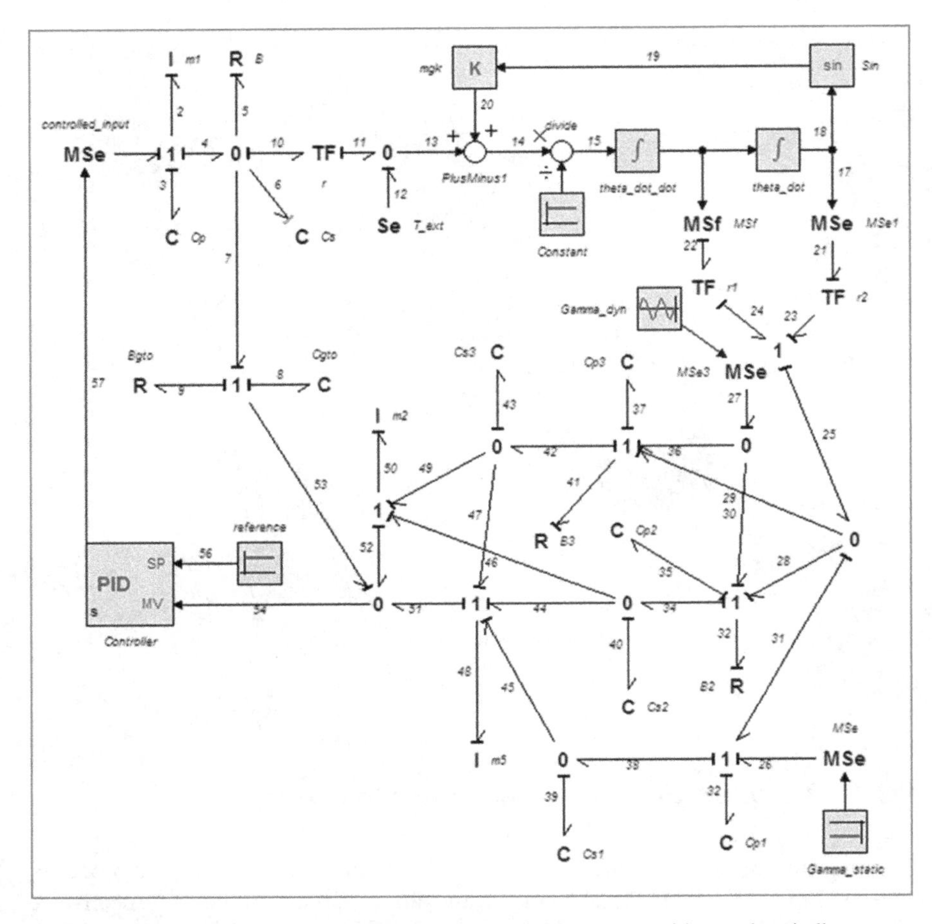

Fig. 9.13 Bond graph model of extended musculoskeletal structure with muscle spindle

9.3 Anthropomorphic Hand through Bond Graph

Anthropomorphic hands are developed for prostheses as well as robotic gripping where human presence is not possible. In addition, such hands are also gaining interest to assist surgeons for tremor free operations [7]. The human hand has a perfect design that gives us ideal movement ability in multiple degrees of freedom as well as grip pattern. Each one of the 27 small bones making up the hand is positioned properly with specific limitations. The muscles that help us to move our fingers are located in our lower arms. These muscles are connected by strong tendon to three small bones in our fingers. There is a special bracelet-like tissue in our wrists that fastens all these tendons. Figure 9.14 shows the block diagram of an anthropomorphic prosthesis hand. The model consists of a DC motor that acts as an

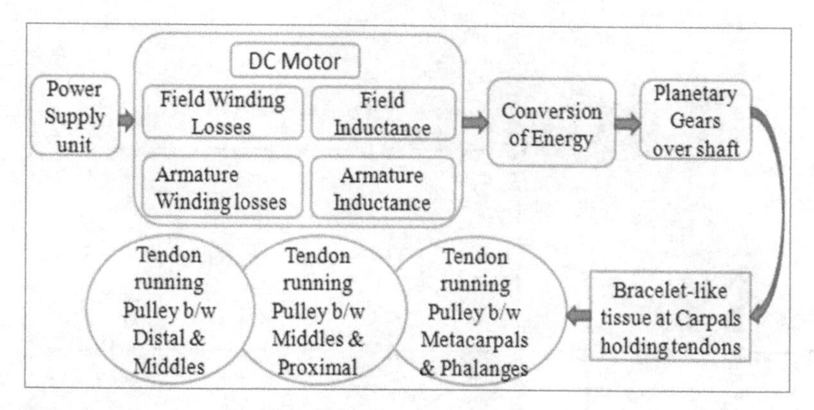

Fig. 9.14 Block diagram of a phalange of an anthropomorphic hand

actuator for the prosthesis. The effect of field and armature winding are modeled to represent the energy dissipation as the winding losses and inductive behavior of winding coil. The electrical energy is converted into mechanical energy as a function of field current. Planetary gear assembly is used to convert the rotational mechanical energy into translational energy, which is the function of 21:13 gear ratio. As the prosthetic mechanism is actuated through DC motor, a string, attached to the shaft of DC motor, is used to carry energy up to each phalange of the prosthetic hand. This string works similarly to the tendons in human hands. Because each finger has different parameters (length, weight etc), the length of each string is different with respect to the finger to which it is attached. These strings are fastened at the shaft just like the bracelet-like tissue that holds the tendons at the carpals in a real hand. The block diagram represents only one phalange, for compactness of block diagram, by three cascaded circles representing proximal, middle and distal extremities. There are five phalanges in the actual model of a hand. The bond graph model is a precise mathematical description of the physical system in the sense that it leads to the state space description of the system.

The bond graph model of an anthropomorphic hand with five phalanges is presented in Fig. 9.15. The bond graph of the anthropomorphic hand has thirteen energy storage elements, and all have integral causalities.

9.4 Movement Coordination Problem for Human Fingers

The anatomy of the hand is efficiently organized to carry out a variety of complex tasks. These tasks require a combination of intricate movements and finely controlled force production [8–9]. The human finger consists of three rigid bodies or bones called phalanges. The phalange closest to the palm is called the proximal interphalange (PIP) followed by the Middle intermediate phalange (MCP), and the

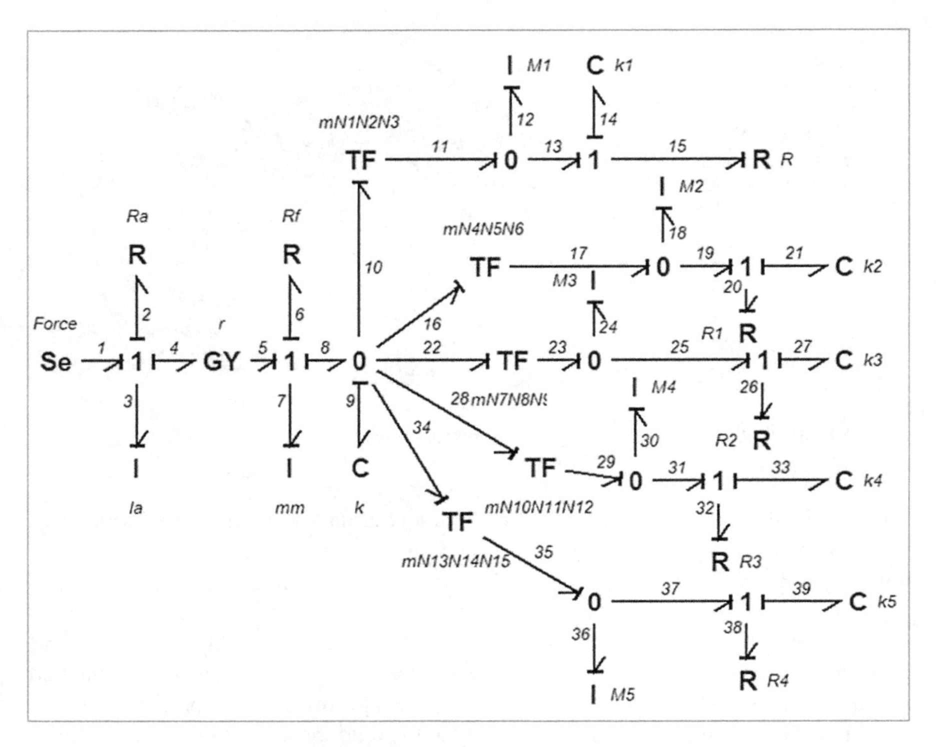

Fig. 9.15 Bond graph model of an anthropomorphic hand of five phalanges implemented with DC motor and pulleys arrangement. The notations used in the bond graph are given as: I_a Inductance of armature winding, R_a armature winding losses, r Gyrator modulus representing the field current, I_{mm} mass of DC motor, R_f frictional losses of motor bearings, k stiffness of the rotating shaft, m gear ratio of planetary gear box, N {1, 4, 7, 10, 13} radius of the pulleys between metacarpals and phalanges of five fingers, N {2, 5, 8, 11, 14} radius of the pulleys between proximal and phalanges of five fingers, N {3, 6, 9, 12, 15} radius of the pulleys between middle and distal of five fingers, M {1 – 5} mass of five phalanges, k {1 – 5} stiffness of string attached to five phalanges, R, R {1 – 4} cumulative damping of pulleys in each phalange

last one, the distal interphalange (DIP). Joints have unique characteristics to provide compliant motion in different directions. The function of tendons is to transmit forces from muscles to bones in order to actuate joints. The tendon lubrication system reduces the friction during tendon excursion. In many cases tendons cross more than one joint and impart rotation to these joints. The extensor mechanism of the fingers contains complex tendinous networks. The nonlinear-elastic approach considers the ligament to respond like a nonlinear spring. By adding a dashpot in series or in parallel to the spring, time-dependent behavior is achieved. The finger kinematic chain consists of three rigid bodies, connected to each other by joints. First, a single rigid body with two connection points for the joints is modeled. Bond graph notation is used to create the model of the rigid body. The actual finger is made such that the phalanges can only rotate around one axis, and the center of gravity can only move in a plane perpendicular to this rotation axis. Muscles provide rotational and translational motion. Joints between the bones are modeled as revolute joints with viscoelastic behavior due to the presence of soft cartilage. Viscoelastic behavior is modeled by providing C and R elements between the 0 junctions of all the bones. The nature of elements C and R depend on the properties of the biological material. Similarly, rotational motion is allowed about one axis only and limited motion is permitted about the other two axes. Consider the distal interphalangeal joint of the little finger: when a force is applied at distal phalanx, a rotational motion is experienced by the phalanx.

The joint tries to oppose rotational motion (bending) beyond a certain limit in order to avoid any injury. That opposition is caused by joint stiffness, similar to stretched spring and friction. Accordingly, a joint can be modeled via a capacitance (C) and resistance (R). Although force has been applied at the distal phalanx of the little finger, it is being transferred to adjacent phalanx through DIP, so a joint performing force transformation also acts as a transformer. The translational momentum of the phalanx can be considered to be concentrated at the center of its mass. Since the motion of the center of gravity of a rigid body, or the rotation with respect to a fixed axis, is defined with reference to an inertial frame, kinetic energy stores directly to the 1-junction representing their velocity or angular velocity. The body's kinetic energy is a function of its velocity. In six-links Biomechanical model, three joints DIP, PIP, & MCP of both little and ring finger will be considered. The complete system with bones' inertia, joints' stiffness, and friction is modeled with the help of bond graph. Five 2- port elements have been used for conversion from translational to rotational motion. Torque is distributed across joints through 2-port elements called transformers and 3-port elements called 0-junction and 1-junction as shown in Fig. 9.16. There are twelve energy storage elements, so the order of the system will be 12; leading to 12 state space equations. Resistances R, are the viscous-elastic damping, Compliances C are stiffness and inertia I are the masses of the corresponding segments.

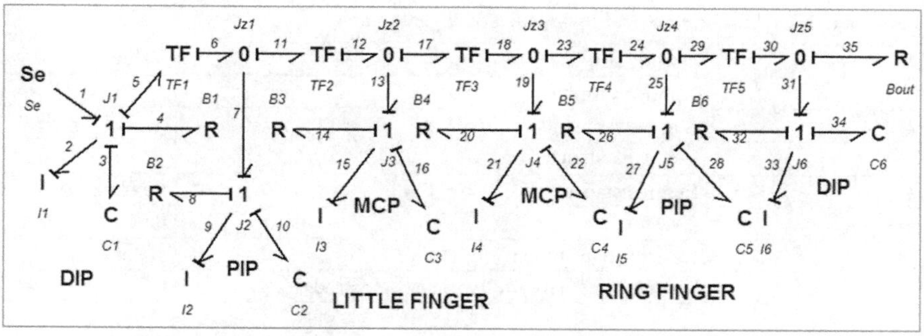

Fig. 9.16 Bond graph model of a movement coordination model of little and ring finger

9.5 Electrical-Impedance Plethysmography

Electrical-impedance plethysmography is used to measure a wide variety of changes in the volume of tissue. Electrodes placed on both legs provide an indication of whether pulsations of volume are normal. If the pulsating waveform in one leg is much smaller than that in the other, this indicates an obstruction in the first leg. If pulsating waveforms are reduced in both legs, this indicates an obstruction in their common supply. A clinically useful noninvasive method for detecting venous thrombosis in the leg is venous-occlusion plethysmography. Electrodes on each side of the thorax provide an excellent indication of the rate of ventilation, but they give a less accurate indication of volume of ventilation. Such transthoracic electrical-impedance monitoring is widely used for infant apnea monitoring to prevent sudden infant death syndrome (SIDS). Electrodes around the neck and around the waist cause current to flow through the major vessels connected to the heart. The resulting changes in impedance provide a rough estimate of beat-by-beat changes in cardiac output. The bond model of electrical-impedance plethysmography is shown in Fig. 9.17. The bond graph is a second-order system. This will give us the value of voltage at the output, by taking value of effort at resistor as described in [10].

The description of variables of Fig. 9.16, is given as Table 9.1.

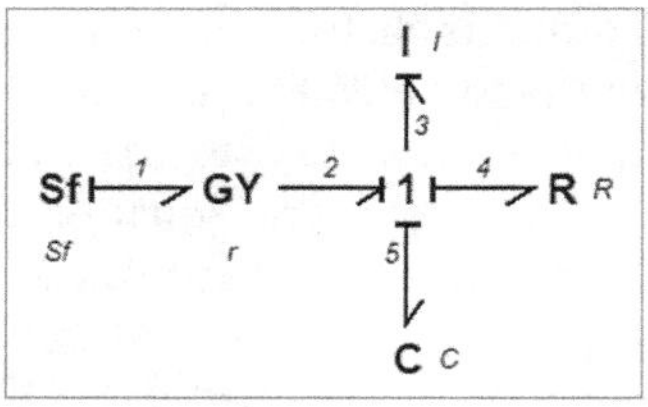

Fig. 9.17 Bond graph model of impedance plethysmography

Table 9.1 Variables of electrical-impedance plethysmography

Variables	Description	Values
Sf	Applied current	1 mA
I_3	Leg inductance	3.6935 kg
C_4	Leg capacitance	10×10^{-6} Farad
R_4	Leg resistance	60 ohm
$r = \rho b L^2$	Gyrator ratio	0.06
ρb	Resistivity of blood	1.5 ohm m
L	The distance between two inner electrodes	0.2 m

9.6 Evaporation in Infant Incubator

Heat transfer mechanisms play an important role in physical systems. There are lots of physical systems in which their study has been conducted in different ways. A baby incubator is a special type of biomedical device that provides an ideal environment for the baby and operates on the mechanism of thermodynamics because premature babies are unable to keep themselves sufficiently warm due to their immature regulation systems. They are also very weak and get infected very quickly. There are four types of thermodynamic processes in baby incubators: conduction, convection, evaporation and radiation. Each plays the important role of keeping the infant warm. There are two types of evaporation that occur in infant incubators. The first evaporation occurs at the surface of the infant due to thermoregulation. To overcome this problem a second type of evaporation is produced in an incubator by introducing air from the outer environment through a mechanical process. This air is warmed with the aid of a heater; water in the tank which is placed in the incubator is heated with the help of this warm air to produce vapors and to warm the inner side of incubator so that the required temperature is provided through latent heat for the infant. Moreover, there is no method for controlling and providing a constant evaporation, which is produced by air coming from the outer environment. We use the pseudo bond graph approach to model the second type of evaporation process in a baby incubator. We show that an incubator is an open loop system and equations for an incubator with the help of the first law of thermodynamics can be obtained to further develop a controller design as shown in [11]. The major cause of hypothermia is the heat loss in small infants that happens due to the evaporation from the skin of infants. In order to overcome this problem outside air is entered into the incubator chamber which is warmed through a heater. With the help of a water tank in the incubator, vapors produce the latent heat which warms the infant body and saves it from this heat loss. These vapors then go to the chamber of the incubator and produce the latent heat which hence provides the required temperature and humidity to the infant. To maintain a constant temperature and humidity in the chamber of the incubator, a microcontroller is placed that takes the information from a temperature and humidity sensor, therefore monitoring as well as controlling the whole mechanism of air and heat flow in the incubator according to the set point. According to Pseudo Bond graph methodology shown in Fig. 9.18 we modeled the air as a source of flow, heating unit (heater) as a heating capacity and processes of coveting water into vapors as a gyrator. Additionally we introduce a capacitance as an addition of water vapors into warmed air, and at the output the heat flow in the chamber is modeled as a source of flow.

This output of this pseudo bond graph model is a value from heat flow sensor which is an input to the controller in order to regulate the air flow (*Sf*).

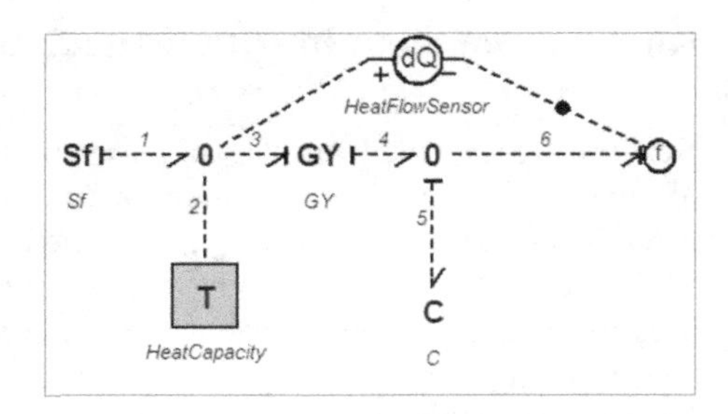

Fig. 9.18 Pseudo bond graph of evaporation

Fig. 9.19 Mechanical structure of the system

9.7 Customized Robotic Arm

Due to the recent advancements in the "Robotics," new and customized robotic parts imitating human body movements are developed everyday. These kinds of developments also help where the presence of human beings is either injurious or deadly. The field of physiotherapy has reached new milestones owing to the introduction of robotic arms and limbs, which replace these amputated parts of a human body [6]. Their need is not only widespread but also desired to make life easier and more comfortable. Their use can be varied to fit our own purposes and requirements. This example discusses modeling of a miniaturized robotic gripper, which moves in a vertical and then horizontal direction to grab any object in consideration [12]. The arm has two aluminum bars at its extreme ends which perform the griping action when instigated. The complete structure on which this arm is placed along with the arm is shown in Fig. 9.19.

The assembly is such that the gripper arm is initially at rest at an initial vertical position, with the two gripping rods of the arm stretched apart. The first motion is the movement of the whole gripper bar from vertical initial point to a vertical final point. After that has been achieved, the bars that perform the gripping action move towards each other. In a modeling approach the whole system in considered as a single plant. A Block diagram of the system is shown in Fig. 9.20. The purpose of

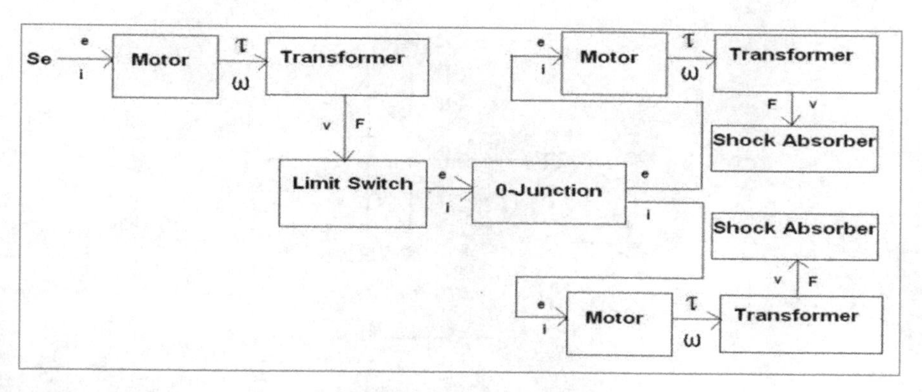

Fig. 9.20 Block diagram of a robotic gripper model

the first motor is to convert the electrical energy through voltage source into rotational mechanical energy. The requirement is to move the bar vertically or translational upwards, so a transformer represents this motion by converting rotational mechanical energy into translational. After the transformation of the energy, this bar reaches the vertical final point and closes a limit switch. This limit switch performs the action of a gyrator because the mechanical energy is converted into electrical energy by relating the velocity (flow) to voltage (effort). After reaching final height, the two gripping rods receive input energy to start moving towards each other, a motion which is again represented by a transformer followed by a shock absorber to dampen any jitters. Motors are represented as gyrators and output is monitored before shock absorbers. The bond graph of the system is given in Fig. 9.21.

9.8 Bilateral Master–Slave Telemanipulation

Telerobotics and telemanipulation refer to the remote operation of a manipulator (usually termed as slave) by slewing it to the movement of a primary manipulator (or master). Realistic telemanipulation involves some kind of kinesthetic feedback to the operator to improve manipulation accuracy. Recent advances in teleoperation include the ideas of collaborative teleoperation. In a collaborative system, a robot also possesses intelligence for secondary-level decision making. Master and slave telemanipulation also provides the basis of collaborative master–slave control, especially in unknown environments, which the human operator does not clearly understand while operating the master. But the actual first step towards collaborative telerobotics is the design of a master–slave system with bilateral controls. Figure 9.22 describes the Master–Slave model for telerobotics. The operators are usually humans who control master devices.

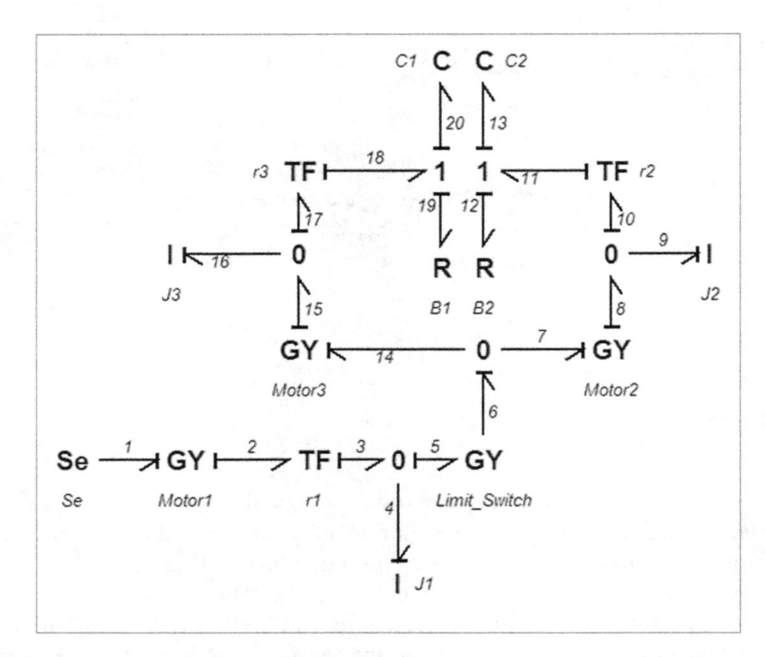

Fig. 9.21 Bond graph of a robotic gripper model of Fig. 9.20

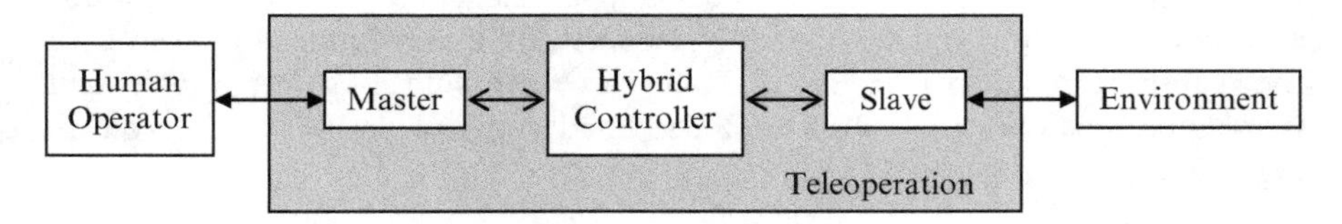

Fig. 9.22 Model for Master–Slave telemanipulation

There is hybrid feedback controller between master and remote slave, which interacts with the environments. The operator commands are usually in the form of force or position input. Similarly, the environment can also exert the force on the slave or change its position. Either combination can also be applied in the entire model and also in the master–slave dynamics. The operator force command is transferred to the slave force command through master and the position commands work similarly. However, force and rate of change of position (velocity) are the variables of interest in our systems and these are described as power variables in the bond graph methodologies. The hybrid model between master and slave is presented in Fig. 9.23. The two-port model presented in Fig. 9.23, can be modeled in various types of hybrid systems depending upon the input and output considerations. From master force in to slave force out system impendence matrix is calculated; similarly velocity in from master to velocity out from slave admittance matrix is to be calculated. Hybrid matrix gives the relationship from master force to

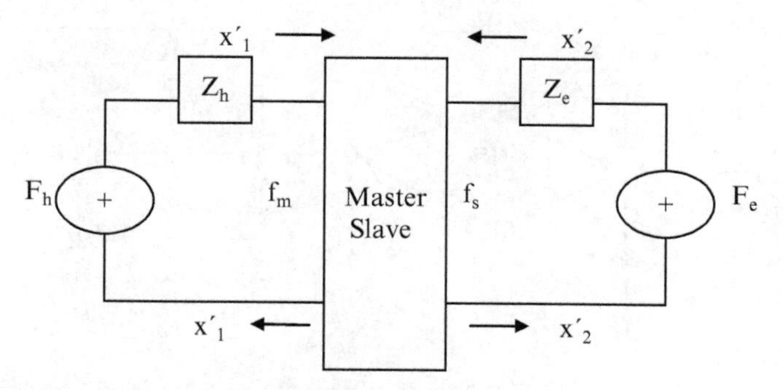

Fig. 9.23 Hybrid model for Master–Slave telemanipulation

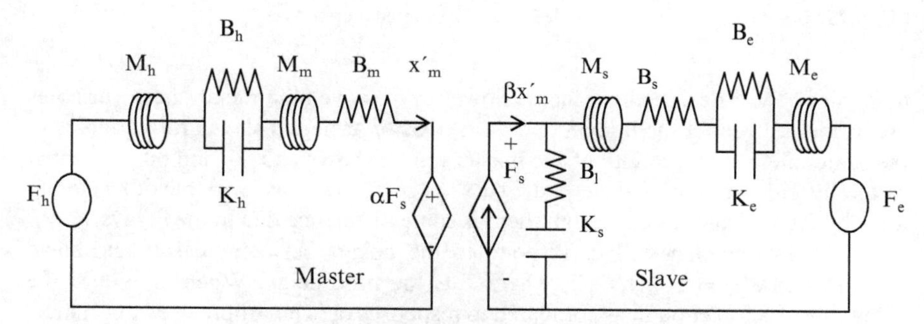

Fig. 9.24 Detailed network model for Master–Slave telemanipulation

the slave velocity and vice versa is the chain matrix. In this paper teleoperation model for complete network presented in [13] is modified by incorporating the human and environmental impedance characteristics as shown in Fig. 9.23. In that figure master–slave configuration is replaced by the dependent voltage and current sources. The voltage-dependent voltage source at master is proportional to the slave force (voltage) and current-dependent current source at slave end is proportional by the velocity (current) at master [14]. F_h is the force delivered by the human operator through the human operator impedance to the master.

The human operator impedance consists of mass M_h of the human operator arm acting on master; B_h and K_h are the viscous and stiffness coefficients of human muscles exerting force. The M_m and B_m are the mass and viscous coefficients of the master arm. Similarly for the slave, there is slave mass M_s, viscous coefficient of slave B_s and viscous coefficient of loss B_l at slave end. K_s represents the stiffness of the slave and B_l is the due to the losses at slave end and these are sometimes negligible. M_e, B_e and K_e are the mass, viscous coefficients and stiffness of the environment or the object slave is interacting with the system. This object at environment also exerts some contact or reaction forces F_e to the slave arm as shown in block diagram in Fig. 9.24. The objects at environment variables have a large range to choose depending upon the specific applications of the robot. If a

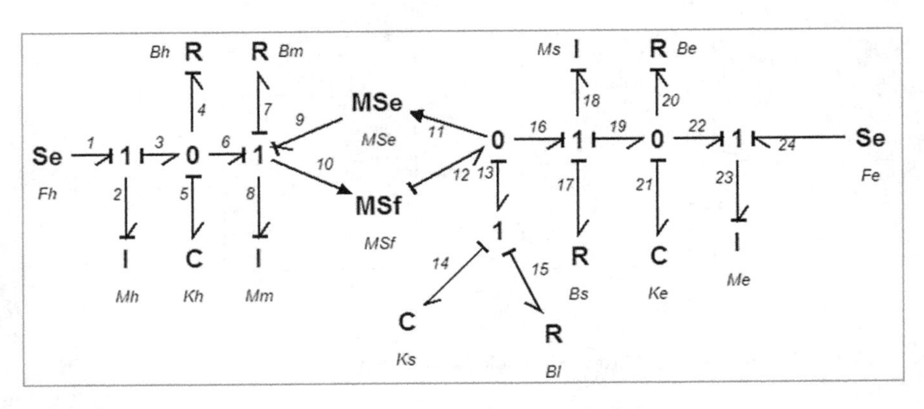

Fig. 9.25 Bond graph model for detailed network of telemanipulation

dynamic robot is designed to pluck a flower or to move high metal, these variables are different keeping the human operators, master arm, and slave arm variables in the same range. The velocity of the master arm is shown as $x'm$ and this is treated as the hybrid variable at slave end *as* $x's = \beta x'm$. The two variables of velocity must be theoretically the same but there is some difference due to the delays. $\beta = \beta(t)$ is a function that will correspond to the delays between master and slave velocities as $x's = x'm(t - T)$, where T is the time delay. When $T = 0$ or the delay time is short enough as compared to responses or is not of precise importance, $\beta = 1$ can be assumed. The force at slave end is controlled and acted through voltage (force) control voltage (force) source. The $F_{\mathrm{m}} = F_{\mathrm{s}}(t - T) = \alpha F_{\mathrm{s}}$ where $\alpha = \alpha(t)$. The slave force is available to the master for transmission but not the contact or reaction force of environmental object, $F_{\mathrm{s}} \approx (1 + \alpha f)F_{\mathrm{e}}$. Therefore, in this paper we have used the total impedance of the environment as $Z_{\mathrm{te}} \approx (1 + \alpha f)Z_{\mathrm{e}}$, where αf is force gain. Figure 9.25 presents a bond graph model from human operators to the environmental object with impedance characteristics. Due to analogies between mechanical and electrical systems, an electrical network similar to the networks is adopted for the bond graph simulation. As the system consists of both electrical and mechanical components, generalized variables are used in the bond graph model. The power variables are "*e*" and "*f*" for effort and flow. Momentum and Displacement are represented as "*p*" and "*q*." The bonds at 10 and 11 are active bonds for effort (voltage or force) and flow (velocity or current). These active bond relations are described in Eq. (9.28).

$$\begin{aligned} e_9 &= \alpha \cdot e_{11} \\ f_9 &= f_{11} = 0 \\ e_{12} &= e_{10} = 0 \\ f_{12} &= \beta \cdot f_{10} \end{aligned} \tag{9.28}$$

This is so because current flow in the 9 and 11 bond remains the same and modulated by an external power source, but effort is changed which is represented by the voltage controlled voltage source, and hence in active bond with source of effort between 9 and 11. Similarly for 10 and 12, the effort remains the same and modulated by an external power source, but flow changes, which is represented by the source of flow between 10 and 12. F_h represents the force exerted by the human operator and F_e represents the contact or reaction force by the object at environment. If the contact force is negligible then SEe = 0 can be assumed, especially when a relative big slave arm is pulling the small object. This assumption will not make any changes in the bond graph. The losses at slave arm are represented by B_l. In case no loss condition is to be required for simulation, bond 14 will be replaced as bond 13 and bond 15 should be deleted.

Problems

P9.1 Simulation of physiological models: Simulate the equations of muscle, muscle spindle, GTO and extended musculoskeletal structures. Use all stiffness values as 99 kN/m, damping values at 1200 N-s/m and masses as 0.1 kg.

P9.2 Control of musculoskeletal structure: Implement the state space equations of extended musculoskeletal structures in MATLAB and control at output response for unit step function. How much effort is needed in bond number 57 in Fig. 9.13 for controlled response?

P9.3 Different models for similar systems: What are the key differences in two models of human-like hands given in Figs. 9.15 and 9.16? How many external control forces are required and what types of hand grip movements are possible?

P9.4 Electrical-Impedance Plethysmography: Use bond graph of Fig. 9.17 and variables from Table 9.1 to obtain state space equations. Simulate these equations in MATLAB with the addition of random white noise (Power 1 μW) at input. The system is stable and this is diagnostic model—how can this problem be extended to develop a better diagnostic measurement by adding linear estimator?

P9.5 Equations of pseudo bond graph: Write state space equations of pseudo bond graph of evaporation problem in Fig. 9.18.

P9.6 Robotic gripper: Explain the working of a robotic gripper through its state space equations. Convert this model to be a feedback control problem as a multivariable control problem.

P9.7 Equations of telemanipulation network: Bond graph of a detailed telemanipulation network is given in Fig. 9.25. Specify state and input vectors by utilizing Eq. (9.28) and verify your results from ref research paper. If only one output is sensed through position sensor at bond number 22 then write output state space equations accordingly and specify whether this position sensor changes the order of state space equations or not. Repeat the problem with both position and velocity sensors at bond number 22 and obtain state space equations. Convert this problem into a feedback control system design with two outputs (i.e., position and velocity at bond number 22).

References

1. Mughal, Asif Mahmood. 2009. *Analytical Biomechanics—Modeling and Optimal Controller Designs*. Saarbrücken: VDM Verlag.
2. Mughal, Asif Mahmood, and Kamran Iqbal. 2013. Bond Graph Modeling and Optimal Control Design for Physiological Motor Control System. *International Journal of Modeling and Simulation* 33(2):93–101.
3. Mughal, Asif M., and Kamran Iqbal. 2006. $H\infty$ Controller Synthesis for a Physiological Motor Control System Modeled with Bond Graphs. *IEEE International Conference of Control Applications*, 947–952, Munich, Germany, 4–6 October.
4. Mughal, Asif M., and Kamran Iqbal. 2006. Modelling and Analysis Physiological Motor Control Using Bond Graph. *IFAC Symposium on Modelling and Control in Biomedical Systems*, Reims, France, 20–22 September.
5. Zoheb, Madiha, and A. Mahmood. 2013. Bond Graph Modeling and PID Controller Stabilization of Single Link Biomechanical Model. *International Conference on Modeling and Simulation*, Islamabad, Pakistan, 25–27 November.
6. Qaisar, Tayyaba, and A. Mahmood. 2014. Robust Control and Optimal Analysis of a Customized Robotic Arm with H_2 and H_∞ Compensators. *The 17th International Multi-Topic Conference*, Karachi, Pakistan, 8–10 December.
7. Furqan, Muhammad, Tallal Saeed, and A. Mahmood. 2013. Anthropomorphic Hand Through Bond Graph Modeling with H_2 and H_∞ Synthesis. *The 16th International Multi-Topic Conference*, Lahore, Pakistan, 19–20 December.
8. Iqbal, Maryam, Mona Jaffer, and A. Mahmood. 2013. Bond Graph Modeling with Control Synthesis of Coordinated Fingers Movement. *International Conference on Modeling and Simulation*, Islamabad, Pakistan, 25–27 November.
9. Iqbal, Maryam, and Mahmood. 2015. H_2 and H_∞ Optimal Control of Coordinated Fingers Movement. *International Conference on Emerging Technologies*, Pakistan, 19–20 December.
10. Javed, Sana, and A. Mahmood. 2013. Analysis on Measurement of Volume of Blood in Human Vessels. *International Conference on Modeling and Simulation*, Islamabad, Pakistan, 25–27 November.
11. Javed, Hasan, and A. Mahmood. 2013. A Study of Thermodynamics and Bond Graph Modelling of Evaporation in Infant Incubator. *9th International Conference on Emerging Technologies*, Islamabad, Pakistan, 9–10 December.
12. Qaisar, Tayyaba, and A. Mahmood. 2015. Robust Control of a Customized Robotic Arm with Unstructured Uncertainties. *International Conference on Emerging Technologies*, Pakistan, 19–20 December.
13. Iqbal, Kamran, and Asif M. Mughal. 2004. A Bond Graph Model of Bilateral Master-Slave Tele-manipulation. *Proceedings of the Fifth IASTED International Conference on Modeling and Simulation*, 362–367, Marina Del Rey, CA, 1–3 March.
14. Mughal, A.M., Q.R. Butt, and A.A. Malik. 2005. Bond Graph Modeling and Simulation of a Piezoelectric Accelerometer System. *Eighth IASTED International Conference on Intelligent Systems and Control*, 155–160, Cambridge, MA, 31 October–2 November.

Index

A

Active bond, 40–42, 58, 59, 61, 165, 182, 183
Actuators, 40, 42, 54, 143, 146, 154, 158, 161, 173
Algebraic loops, 61, 85–87, 89, 100
Amplification, 42, 100, 110, 118–119, 123, 126, 134
Analytical modeling., 4–6
Anthropomorphic hand, 172–174
Autonomous system model, 5

B

Bilateral, 179–183
Bridge circuit, 96, 100, 101

C

Capacitor bank, 100, 101
Cartesian, 22–24, 26–28, 32
Causality, 4, 40–42, 48–62, 64, 66, 67, 69, 70, 73–74, 82, 85, 87–91, 95, 96, 98, 100–102, 169, 173
Combination of elements, 82, 96–97
Compensation, 145, 152, 154
Competence, 3
Complex poles, 121, 139
Computational systems, 6–7
Constitutive equations, 59–61, 91
Control design, 150–154
Controller, 78, 82, 118, 143–146, 148, 152–157, 169, 177, 180
Control system, 2, 79, 125, 143–158, 183
Critically, 113–114, 116, 118, 120, 125–133, 140

D

Damped, 109–117, 119, 121–123, 126, 129–133, 140, 151, 152
Decaying, 48, 107, 116, 124, 131–134, 137–139, 152
Derivative, 7, 9, 19, 23, 24, 31, 49, 51–52, 83, 87–90, 153, 154
Discrete systems, 6
Distributed model, 5
Dynamical system, 1–17, 35

E

Eigenvalues, 105–108, 110, 112–118, 121, 124, 131, 134–136, 138–141
Electromechanical network, 72–74, 78, 79
Energy variables, 26, 33–37, 42, 45, 49, 162
Error model, 146–147
Estimator, 147–150, 155, 183
Evaporation, 177–178, 183
Excitation, 36, 39, 40, 104, 121, 124, 128–133, 138, 150, 164, 165, 169
Exogenous inputs, 144–145
Experimental validation, 4
External stability, 134–137

F

Fields, 9, 20, 21, 24, 27, 45, 85, 90–97, 101, 103, 161, 173, 174, 178
Forced, 104, 123–129, 134–139

A.M. Mughal, *Real Time Modeling, Simulation and Control of Dynamical Systems*,
DOI 10.1007/978-3-319-33906-1

G

Gain scheduling, 157
Golgi tendon organ (GTO), 83, 167–169
Gyrator, 52, 54, 55, 58–59, 64, 72, 73, 78, 90, 92, 96, 97, 100, 174, 176, 177, 179

H

Harmonic oscillator, 139, 140
Hinge movement, 60
Human fingers, 173–175

I

Impulse, 39, 124–127, 137
Inertial, 162, 163, 167–169, 175
Infant incubator, 177–178
Integral causality, 50–52, 64, 66, 69, 72, 73, 173
Internal stability, 121–123, 134–136
Inverted pendulum, 29, 30, 32, 146, 167, 168

J

0-Junction, 54–58, 63–66, 68, 69, 72–75, 80, 88, 92, 96, 169, 175
1-Junction, 54–58, 63–66, 68, 69, 71–73, 75, 77, 78, 80, 85, 87, 88, 96, 175
Junctions, 47, 54–58, 63–66, 68, 69, 71–75, 77, 78, 80–82, 85, 87, 88, 92, 96, 98, 162, 169, 175

L

Lagrangian, 19–33, 63, 161
Law of conversation of energy, 20, 54
Linearization, 10–12, 16, 17, 82–84, 147
Lumped system model, 5

M

Marginally stable, 121, 141
Master–Slave, 179–183
Movement coordination, 173–175
Multi-energy system, 63
Multivariable control, 157, 183
Muscle spindle, 164–169, 171, 172
Muscular structures, 162–164
Musculoskeletal, 162, 164, 168–172, 183

N

Natural frequency, 109, 111, 112, 117, 129, 131, 139, 140, 151, 152
Non-causal system, 6
Nonlinear systems, 6, 9, 15, 17, 32, 147, 168

O

Observer, 147–150, 154, 155, 158
Oscillations, 109–113, 115, 116, 123, 133
Outputs, 3, 6–9, 11–17, 39–42, 56–59, 61, 68, 70, 81, 82, 84, 89, 93, 95, 100, 104–107, 121, 124, 125, 127, 129, 134–136, 143–145, 147–150, 153, 154, 157, 163, 164, 166–169, 176, 177, 179, 180, 183
Overdamped, 111–114, 116–118, 120, 125–133, 138
Over Shoot, 155

P

Performance specifications, 150–152, 155, 157, 158
Physiological elements, 161–168
PID tuning, 155, 158, 171
Piston and cylinder, 60
Plethysmography, 176, 183
Pole placement, 155, 158
1-Port capacitor, 49–50, 74
1-Port elements, 47–51, 54, 71, 90, 97
2-Port elements, 52–54, 59, 61, 63, 64, 175
1-Port inertia, 50–51, 75
1-Port resistor, 48–49
Power variables, 35–40, 48–51, 180, 182
Proportional–integral and derivative (PID) control, 73, 146, 150, 152–154, 156, 157
Pseudo bond graph, 37, 177, 178, 183

R

Redundancy, 4, 82, 169
Redundant systems, 4, 17, 52, 88
Reference inputs, 143–146, 149
Resistive, 92, 95–96, 101
Resonance, 129–131
Response, 7, 35, 39, 70, 103–141, 143, 150–156, 169, 182, 183
Rise time, 151, 152, 158
Robotic gripper, 172, 178–180, 183
Rotating pendulum, 32

S

Second order system, 31, 85, 103, 107–115, 117–121, 129, 138, 150–152, 154
Sensors, 42, 43, 45, 54, 61, 70, 72, 73, 82, 143–147, 150, 158, 161, 164, 165, 177, 183
Series motor, 93–94, 100
Settling time, 131, 151, 152
20-Sim, 70, 82–84, 91, 92, 162
Simple pendulum, 28, 146
Simplicity, 3, 97
Sinusoidal, 48, 110, 115, 116, 123, 124, 128, 129, 131–138
Sources, 35, 39–43, 45, 47–48, 51, 59–62, 64, 66–69, 71–75, 78, 84, 87, 89–91, 94, 96, 97, 100, 104, 144, 149, 162, 164, 165, 167, 177, 179, 181–183
Spherical, 22–24, 32
Spring, 50, 70–73, 76, 77, 79, 88, 90, 92, 93, 97, 128, 139, 161, 164, 167, 174, 175
Stable systems, 119, 121, 156
State space method, 8–9
State variables, 2, 8–10, 14, 16, 33, 52, 58, 74–76, 81, 88, 89, 92, 146, 163, 165, 168, 169
State vector, 2, 5, 8, 9, 28, 29, 32, 74, 78, 83, 98, 99, 104, 145–147
Steady state value, 116, 128, 150–152
Step, 39, 43, 51, 64, 82, 124, 127–128, 155, 156, 179, 183
Systems engineering, 1, 2

T

Telemanipulation, 179–183
Thermodynamic, 177
Time varying/time invariant system, 5–7
Total, 7, 14, 20, 30, 31, 54, 103, 105, 137–139, 163, 182
Transfer function, 12–16, 19, 39, 84, 100, 105, 107, 109, 112–114, 117, 124, 127, 129, 135, 136, 140, 143, 145, 146, 149, 150, 153, 154, 158
Transformer, 41, 52–54, 58–60, 64, 68, 69, 73, 78, 87, 90–93, 96, 97, 100, 102, 168, 169, 175, 179

U

Undamped systems, 109–110, 114, 116, 117, 120, 128–129, 131, 138, 139
Under damped, 111, 114–116, 118, 120, 125–133, 140
Undershoot, 112, 151, 152, 156
Unstable systems, 119, 121, 126, 129, 156

V

Vector bond graphs, 47, 97–102, 168

W

Word bond graph, 39–43, 45, 60, 64
Work energy principle, 19

CPSIA information can be obtained
at www.ICGtesting.com
Printed in the USA
LVHW05s2322210618
581519LV00004B/19/P